PROBLEM BOOK IN THE THEORY OF FUNCTIONS

VOLUME II

Problems in the Advanced Theory of Functions

By DR. KONRAD KNOPP

Professor of Mathematics, University of Tübingen

Translated by FREDERICK BAGEMIHL

DOVER PUBLICATIONS, INC.

NEW YORK

Published in Canada by General Publishing Company, Ltd., 30 Lesmill Road, Don Mills, Toronto, Ontario.

Published in the United Kingdom by Constable and Company, Ltd., 10 Orange Street, London WC2H 7EG.

This Dover edition, first published in 1952, is a new English translation of the second edition of *Aufgabensammlung zur Funktionentheorie,* II Teil.

International Standard Book Number: 0-486-60159-5
Library of Congress Catalog Card Number: 49-7400

Manufactured in the United States of America
Dover Publications, Inc.
180 Varick Street
New York, N. Y. 10014

CONTENTS

[1] See the Foreword.

FOREWORD

This translation is of the fourth (1949) edition of K. Knopp's *Aufgabensammlung zur Funktionentheorie, II. Teil.*

References to the first volume of this Problem Book are indicated by writing "I" together with chapter, page or paragraph, and problem numbers; references to the present second volume contain only page or paragraph, and problem numbers. This second volume adheres strictly to the little function-theoretical volumes put out (except the last one) by Dover Publications. These works are quoted as follows:

I	=	*Problem Book in the Theory of Functions*, vol. I: *Problems in the Elementary Theory of Functions*, translated by L. Bers, 1948.
Elem.	=	Knopp, *Elements of the Theory of Functions*, translated by F. Bagemihl, 1952.
K I	=	Knopp, *Theory of Functions*, Part I, translated by F. Bagemihl, 1945.
K II	=	Knopp, *Theory of Functions*, Part II, translated by F. Bagemihl, 1947.
B	=	Bieberbach, *Einführung in die konforme Abbildung*, 4th. ed., Berlin, 1949.

The present volume again contains mainly exercises which remain essentially within the range of ideas appearing in the volumes just mentioned. Only within this frame, and not in any absolute sense, is the division into elementary and advanced function theory intended. The present problems are based mostly on the last chapters of K I, as well as on K II and B. For using the book, the prefatory remarks to I are pertinent. Corresponding to the higher level now, the asterisk (*) is used more sparingly to denote the harder problems.

PART I – PROBLEMS

CHAPTER I

ADDITIONAL PROBLEMS FOR I, CHS. 1–5

§ 1. *Fundamental Concepts*

1. Let k points $z_1, z_2, \ldots, z_k$ be given in the plane of complex numbers, and let $\alpha_1, \alpha_2, \ldots, \alpha_k$ be nonnegative numbers, for which $\alpha_1+\alpha_2+\cdots+\alpha_k = 1$. Then the number $\zeta = \alpha_1 z_1+\alpha_2 z_2+\cdots+\alpha_k z_k$ lies in the smallest convex (closed) polygon containing the points $z_1, z_2, \ldots, z_k$.— Is this still true for infinite sequences $\{z_n\}$ and $\{\alpha_n\}$, provided that $\sum\alpha_n = 1$ and $\sum\alpha_n z_n = \zeta$ exists?

2. If the z_ν and α_ν have the same meaning as in the preceding problem, then every point ζ for which

$$\frac{\alpha_1}{\zeta - z_1} + \frac{\alpha_2}{\zeta - z_2} + \cdots + \frac{\alpha_k}{\zeta - z_k} = 0,$$

lies in the smallest convex (closed) polygon enclosing the points $z_1, z_2, \ldots, z_k$.

3. The smallest convex (closed) polygon enclosing the roots of an entire rational function $G(z)$ also contains all the roots of its derivative $G'(z)$.

*4. Let $n \geqq 4$ points be given in the plane of complex numbers, no four lying on a circle. Let $l_\nu(z)$ be a linear transformation for which $l_\nu(z_\nu) = \infty$, and let the smallest convex polygon π_ν containing the images of the remaining points have a_ν vertices, $\nu = 1, 2, \ldots, n$. Then the following assertions are true:

a) $a_1+a_2+\cdots+a_n = 6\ (n-2)$,

b) not all the π_ν have 6 or more vertices,

c) for $n<12$ a triangle or a quadrilateral always appears among the π_ν,

d) if, for $n \geqq 12$, the triangles and quadrilaterals are lacking entirely, then there are at least 12 pentagons among the π_ν.

*5. Let $n \geqq 4$ points be given in the plane of complex numbers, no four lying on a circle. They then determine $\binom{n}{3}$ distinct circles. Precisely $2\,(n - 2)$ of these circles have the property that one of the two regions into which the circle divides the plane contains none of the given points.

6. Every closed polygon $\mathfrak{p}$ which intersects itself arbitrarily (i.e., every curve which is composed of any finite number of arbitrary, oriented straight line segments joined in order in such a manner that the initial point of each coincides with the terminal point of the preceding segment, and such that the last point of the curve coincides with the first) can be decomposed into a finite number of simple closed polygons and a finite number of segments described twice, once in each direction. Each of the former is described either entirely in the positive or entirely in the negative sense.

§ 2. *Infinite Sequences and Series*

1. The convergence of both series

$$\sum_{n=0}^{\infty} a_n \quad \text{and} \quad \sum_{n=0}^{\infty} |b_n - b_{n+1}|$$

implies that, for every positive integer q, the series

$$\sum_{n=0}^{\infty} a_n\, b_n^q.$$

converges.

2. From the two infinite series $\sum_{n=0}^{\infty} a_n$ and $\sum_{n=0}^{\infty} b_n$ form a third series $\sum_{n=0}^{\infty} c_n$ according to the rule

$$c_n = a_0 b_n + a_1 b_{n-1} + \cdots + a_n b_0.$$

Then the following three theorems hold:

a) If the first two series converge absolutely, and have the respective sums A and B, then the third series also converges, and has the sum $C = AB$.

b) For the validity of the conclusion of Theorem a), the absolute convergence of only one of the first two series, and at least the conditional convergence of the other, is sufficient.

c) If all three series converge (even if only conditionally), then their sums A, B, C satisfy the relation $AB = C$.

3. If $\sum_{n=0}^{\infty} a_n$ converges, then the two series

$$\text{a)}\ a_0 + \sum_{n=1}^{\infty} \frac{a_1 + 2a_2 + \cdots + na_n}{n(n+1)}$$

$$\text{b)}\ \sum_{n=0}^{\infty} \frac{1}{2^{n+1}} \left[\binom{n}{0} a_0 + \binom{n}{1} a_1 + \cdots + \binom{n}{n} a_n \right]$$

also converge and have the same sum as $\sum_{n=0}^{\infty} a_n$.

4. The series $\sum_{n=1}^{\infty} \frac{1}{n^{1+i\alpha}}$ converges for *no* real α.

5. The series $\sum_{n=2}^{\infty} \frac{1}{n^{1+i\alpha} \log n}$ converges for *every* real $\alpha \neq 0$.

6. For what values of z is the equation

$$\sum_{n=0}^{\infty} (-1)^n \binom{z}{n} = 0,$$

obtained by formally expanding $(1-1)^z$, valid?

*7. Show that the following identities hold for $|z|<1$:

$$\text{a)}\ \sum_{n=1}^{\infty} (-1)^{n-1} \frac{z^n}{1-z^n} \equiv \sum_{n=1}^{\infty} \frac{z^n}{1+z^n},$$

$$\text{b)}\ \sum_{n=1}^{\infty} n \frac{z^n}{1-z^n} \equiv \sum_{n=1}^{\infty} \frac{z^n}{(1-z^n)^2},$$

$$\text{c)}\ \sum_{n=1}^{\infty} \frac{(-1)^{n-1}}{n} \frac{z^n}{1-z^n} \equiv \sum_{n=1}^{\infty} \log(1+z^n),$$

$$\text{d)}\ \sum_{n=1}^{\infty} \alpha^n \frac{z^n}{1-z^n} \equiv \sum_{n=1}^{\infty} \frac{\alpha z^n}{1-\alpha z^n}, \quad |\alpha| < 1.$$

*8. Let $f(z) = \sum_{n=0}^{\infty} a_n z^n$ be a power series with radius of convergence $r > 0$, (b_n) a sequence of numbers such that

$$\frac{b_n}{b_{n+1}} \to \beta,$$

and $|\beta| < r$. Then

$$\frac{c_n}{b_n} = \frac{a_0 b_n + a_1 b_{n-1} + \cdots + a_n b_0}{b_n} \to f(\beta).$$

*9. Let α and β be two positive numbers whose sum is 1. For a given z_0, with $\Re(z_0) \lessgtr 0$, set

$$z_1 = \alpha z_0 + \frac{\beta}{z_0},$$

so that we have also $\Re(z_1) \lessgtr 0$; and in general, for $n \geqq 1$, put

$$z_n = \alpha z_{n-1} + \frac{\beta}{z_{n-1}},$$

so that invariably $\Re(z_n) \lessgtr 0$. Then

$$z_n \to +1, \text{ if } \Re(z_0) > 0,$$
$$z_n \to -1, \text{ if } \Re(z_0) < 0.$$

For $\Re(z_0) = 0$ the sequence is either meaningless or divergent.

10. In connection with § 1, Prob. 1 and I, § 3, Prob. 6, show the following: Let $\{z_n\}$ be an arbitrary bounded sequence of numbers and let $\mathfrak{K}$ be the smallest convex domain containing all its limit points ζ. Then if the sequence $\{z_n'\}$ is derived from the sequence $\{z_n\}$ as in I, § 3, Prob. 6, and if all the transformation coefficients $a_{\kappa\lambda} \geqq 0$, then $\mathfrak{K}$ also contains all limit points ζ' of the sequence $\{z_n'\}$.

§ 3. *Functions of a Complex Variable*

1. a) Can a function of a complex variable which is defined and continuous for $|z| < 1$ be such that it is only differentiable at the origin? b) Can a function $f(z)$ which is continuous in a region be differentiable only along certain lines in that region?

2. Can a function $f(z)$ defined in a region $\mathfrak{G}$ be differentiable at each point of an everywhere dense set of points in $\mathfrak{G}$ and at the same time be *non*differentiable at each point of another everywhere dense set of points in $\mathfrak{G}$?

3. Can a function $f(z)$ which is defined and differentiable in a region $\mathfrak{G}$ be everywhere real or, more generally, have a constant real part, or a constant imaginary part, or a constant amplitude, or a constant absolute value, in a subregion $\mathfrak{G}_1$ of $\mathfrak{G}$?.

4. Let the function $f(z)$ be differentiable in $|z|<1$, and let its derivative $f'(z)$ be bounded there. Then $f(z)$ assumes continuous boundary values along $|z| = 1$, and these together with $f(z)$ form a uniformly continuous function in $|z| \leqq 1$. (Cf. I, § 5, Probs. 8 and 9.)

5. $\sqrt{x+iy} = \pm\left\{\sqrt{\tfrac{1}{2}(\sqrt{x^2+y^2}+x)} + i\sqrt{\tfrac{1}{2}(\sqrt{x^2+y^2}-x)}\right\}$, if the nonnegative value is taken for $\sqrt{x^2+y^2}$, and if for $y \lesseqgtr 0$ the signs of the two (outer) radicals in the braces are chosen so that the product of these roots has the same sign as y.

6. What inconsistency with other agreements concerning powers lies in the fact that e^z is in general understood to be the sum of the series $\sum_{n=0}^{\infty} \frac{z^n}{n!}$?

7. a) What is the limit of the function $\exp[1/(1-z)]^2$ if z approaches the point $+1$ rectilinearly from the interior of the unit circle?

b) How does $\exp[1/(1-z)]^2$ behave if z approaches $+1$ along the circumference?

8. a) The function z^a is single-valued if, and only if, a is a real integer.

b) For all z of the plane, the absolute value of the principal value of the function z^i remains less than a fixed constant.

9. The branch of the function $f(z) = (1-z)^i$ which is regular for $|z|<1$ and equals $+1$ at $z=0$ has an absolute value which, for $|z|<1$, lies between two fixed positive numbers.

*10. How does the function $z^3 \exp(-i \operatorname{Log}^2 z)$ behave, where $\operatorname{Log} z$ denotes the principal value, if z approaches 0 rectilinearly from the (closed) right half-plane?

§ 4. *Integral Theorems*

1. a) Can a rectifiable path be such that it possesses *no* tangent at one or at infinitely many of its points?

*b) Can it possess *no* tangent at points that are everywhere dense on the path, or even at all points of the path?

*2. Let the rectifiable path $\mathfrak{k}$ be given by the parametric representation

$$x = \varphi(t),\ y = \psi(t),\ \alpha \leqq t \leqq \beta.$$

Let the functions $f(z)$ and $\rho(z)$ be continuous along $\mathfrak{k}$, where $\rho(z)$, moreover, is assumed to be real and positive. Let $\mathfrak{k}'$ be the image of $\mathfrak{k}$ under the mapping given by $w = f(z)$. Then, if we set

$$\int_{\alpha}^{\beta} f(z)\rho(z)dt = \mu \cdot \int_{\alpha}^{\beta} \rho(z)dt,$$

μ lies in every convex polygon containing the points of $\mathfrak{k}'$.

3. Let $F(z)$ be a continuous function of z along the (closed) path $\mathfrak{k}$. Show that the limiting value

$$\lim_{n\to\infty} \left\{ \sum_{\nu=1}^{n} |z_\nu - z_{\nu-1}| \cdot F(\zeta_\nu) \right\} = \int_{\mathfrak{k}} F(z)\,|dz|,$$

understood in the usual sense, always exists (cf. K I, § 9).

4. If $f(z)$ is a continuous function of z along the path $\mathfrak{k}$, then invariably

$$\left| \int_{\mathfrak{k}} f(z)dz \right| \leqq \int_{\mathfrak{k}} |f(z)|\,|dz|,$$

if the integral on the right is understood in the sense of the preceding problem (with $F(z) = |f(z)|$).

5. What is the connection between the integral defined in Prob. 3 and the line integrals

$$\int_{\mathfrak{k}} U(x, y)ds \quad \text{and} \quad \int_{\mathfrak{k}} V(x, y)ds,$$

if we set $F(z) = U(x, y) + iV(x, y)$?

6. Prove the theorem formulated in I, § 5, Prob. 13, without separation into real and imaginary parts, by using Cauchy's integral formula.

*7. Let the function $F(z, t)$ be defined whenever z lies in a certain region $\mathfrak{G}$, and t at the same time lies on a path $\mathfrak{k}$ in the z-plane, extending from a to b. For all these pairs of values let $F(z, t)$ be bounded, say $|F(z, t)| < M$.

Then, if

a) for every fixed t on $\mathfrak{k}$, $F(z, t)$ is a regular function of z in $\mathfrak{G}$, and

b) for every fixed z in $\mathfrak{G}$, $F(z, t)$ is a continuous function of t along $\mathfrak{k}$,

the integral

$$\int_{\mathfrak{k}\,a}^{b} F(z, t)\,dt = f(z)$$

defines a regular function of z in $\mathfrak{G}$, whose derivative is given by

$$f'(z) = \int_{\mathfrak{k}\,a}^{b} F_z(z, t)\,dt.$$

(Hint: Consider the integral

$$\frac{1}{2\pi i} \int_{\mathfrak{k}\,a}^{b} \left\{ \int_{\mathfrak{K}} \frac{F(\zeta, t)}{\zeta - z}\, d\zeta \right\} dt,$$

where $\mathfrak{K}$ is the circumference of a small circle about z, and apply the methods in K I, § 16).

§ 5. *Expansions in Series*

1. Let $\mathfrak{G}$ be a closed, bounded region. Let the series $\sum_{n=0}^{\infty} f_n(z)$ be uniformly convergent in that part of a certain neighborhood of every point z_0 of $\mathfrak{G}$, which belongs to $\mathfrak{G}$. Then the series is uniformly convergent in $\mathfrak{G}$.

2. Prove Theorem 3 in K I, § 19 without the help of Morera's theorem, say on the basis of Cauchy's integral formula and the theorems in K I, § 16.

3. In connection with K I, § 19, Theorem 3, show that if, in addition to the series $\sum f_n(z)$, the series $\sum |f_n(z)|$ converges uniformly in every $\mathfrak{G}'$, then this Theorem 3 can be sharpened to the effect that

the series $\sum |f_n^{(p)}(z)|$, for fixed p, also converge uniformly in every $\mathfrak{G}'$.

4. Let the functions $f_0(z), f_1(z), \ldots$ be regular in $\mathfrak{G}$, and suppose that $\sum_{n=0}^{\infty} f_n(z) = F(z)$ is uniformly convergent in every closed subregion of $\mathfrak{G}$, but $F(z)$ is not identically 0. Then the zeros of $F(z)$ in $\mathfrak{G}$ are identical with the limit points in $\mathfrak{G}$ of the zeros of the partial sums

$$s_n(z) = f_0(z) + \cdots + f_n(z).$$

5. The power series

$$f(z) = (1 - z)^{-i} = \sum_{n=0}^{\infty} (-1)^n \binom{-i}{n} z^n \equiv \sum_{n=0}^{\infty} b_n z^n$$

converges for $|z| < 1$, and $|f(z)| < e^{\pi/2}$ there. Show, further, that

$$|nb_n| \to \left(\frac{e^{\pi} - e^{-\pi}}{2\pi}\right)^{\frac{1}{2}}.$$

6. The principal value of the function $(1+z)^{1/z}$ is regular in a neighborhood of the origin. Obtain the first few terms of the corresponding power-series expansion.

*7. In connection with I, § 10, Prob. 1a, set up the complete power-series expansion of the function

$$f(z) = e^{\alpha\frac{z}{1-z}} = \sum_{n=0}^{\infty} c_n z^n,$$

and show that for positive α the coefficients c_n admit of an estimate of the form

$$c_n = e^{2\sqrt{\alpha n}\,(1 + \varepsilon_n)},$$

where the ε_n form a null sequence.

CHAPTER II

SINGULARITIES

§ 6. *The Laurent Expansion*

1. Let the simple closed path $\mathfrak{K}_2$ lie in the region bounded by the simple closed path $\mathfrak{K}_1$. Let $f(z)$ be single-valued and regular in the annular region between the two paths. Then one can set

$$f(z) = f_1(z) + f_2(z),$$

where $f_1(z)$ denotes a function which is regular inside $\mathfrak{K}_1$ and $f_2(z)$ denotes a function which is regular in the exterior of $\mathfrak{K}_2$ (including ∞). By this decomposition of $f(z)$, moreover, the functions $f_1(z)$ and $f_2(z)$ are uniquely determined to within an additive constant.

2. Extend the theorem of the preceding problem to the case in which $\mathfrak{K}_1$ encloses several simple closed paths $\mathfrak{K}_2, \mathfrak{K}_3, \ldots, \mathfrak{K}_m$ $(m>2)$ which neither intersect nor enclose one another, and $f(z)$ is now assumed to be single-valued and regular in the region lying inside $\mathfrak{K}_1$ but outside each of the paths $\mathfrak{K}_2, \ldots, \mathfrak{K}_m$.

3. Expand the functions

a) $e^{\frac{1}{z-1}}$ for $|z|>1$,

b) $\sqrt{(z-1)(z-2)}$ for $|z|>2$,

c) $\dfrac{1}{(z-a)(z-b)}$ for $0<|a|<|z|<|b|$,

d) the same function for $|z|>b$,

e) $\log \dfrac{1}{1-z}$ for $|z|>1$

in their Laurent series.

4. Let

$$g(z) = \sum_{n=0}^{\infty} a_n z^n \text{ be an entire function of } z,$$

$$\gamma(z) = \sum_{n=0}^{\infty} \frac{\alpha_n}{z^n} \text{ be an entire function of } \frac{1}{z}.$$

Where is the function $g(z) \cdot \gamma(z)$ regular, and what is its Laurent expansion there?

5. Let $\sum_{n=-\infty}^{+\infty} a_n z^n$ and $\sum_{n=-\infty}^{+\infty} b_n z^n$ be two Laurent expansions which converge in the same circular ring. What is the Laurent expansion of the product of the functions represented by them?

6. What is the Laurent expansion of the function

$$e^{c\left(z+\frac{1}{z}\right)}$$

which is regular in $0<|z|<+\infty$?

§ 7. *The Various Types of Singularities*

1. What kind of singularity does the function

a) $\dfrac{z^2+4}{e^z}$, b) $\sqrt{(z-1)(z-2)}$, c) $\cos z - \sin z$,

d) $\dfrac{1}{\cos z}$, e) $\cot z$, f) $e^{-\frac{1}{z^2}}$,

g) $\sin \dfrac{1}{1-z}$, h) $\dfrac{1}{1-e^z}$, i) $\dfrac{1-e^z}{1+e^z}$,

have at $z = \infty$?

2. What sort of singularity has the function

a) $e^{\frac{1}{z}}$ at $z = 0$, b) $\sin \dfrac{1}{1-z}$ at $z = 1$,

c) $\dfrac{1}{1-e^z}$ at $z = 2\pi i$, d) $\dfrac{1}{\sin z - \cos z}$ at $z = \dfrac{\pi}{4}$?

3. At the point z_0, let the function $f_1(z)$ have a zero of order α, and the function $f_2(z)$ have a pole of order β ($\alpha>0$, $\beta>0$). What sort of point is z_0 for

$$f_1 \pm f_2, \quad f_1 \cdot f_2, \quad \frac{f_1}{f_2}, \quad \frac{f_2}{f_1}?$$

4. With the aid of Theorem 2 in K I, § 33, prove that e^z has no zeros.

5. What essential difference is there between the behavior of the real function

$$y = \begin{cases} e^{-\frac{1}{x^2}} & \text{for } x \neq 0, \\ 0 & \text{for } x = 0 \end{cases}$$

and that of the function of a complex variable $w = e^{-\frac{1}{z^2}}$ in a neighborhood of the origin?

6. Let the function $f(z)$ have a pole of order β ($\beta \geqq 1$) at z_0, and be regular everywhere else in the circle $\mathfrak{K}$ about z_0. Under what conditions are the functions

$$F_0(z) = \int_{z_0 \atop \mathfrak{k}}^{z} f(z)dz \quad \text{and} \quad F_1(z) = \int_{z_1 \atop \mathfrak{k}}^{z} f(z)dt$$

$$(\mathfrak{k} \text{ and } z_1 \neq z_0 \text{ in } \mathfrak{K})$$

single-valued and regular in a neighborhood of z_0, and how do they behave at z_0 in such a case?

7. Let $\mathfrak{k}$ be an arbitrary bounded path (closed or open), and let the function $\varphi(z)$ be defined and continuous along $\mathfrak{k}$. Then (see K I, § 16)

$$f(z) = \frac{1}{2\pi i}\int_{\mathfrak{k}} \frac{\varphi(\zeta)}{\zeta - z}\, d\zeta$$

is single-valued and regular outside a sufficiently large circle $\mathfrak{K}$ about 0. How does $f(z)$ behave at $z = \infty$?

8. Let $f(z)$ be single-valued for $|z| > R$, and, except perhaps at $z = \infty$, also regular. Under what conditions is the function

$$F(z) = \int_{z_0 \atop \mathfrak{k}}^{z} f(\zeta)d\zeta$$

also single-valued and regular there, if z_0 and $\mathfrak{k}$ lie in $|z| > R$? What does the behavior of $f(z)$ at ∞ imply concerning that of $F(z)$ at ∞?

9. Prove Riemann's theorem, K I, § 32, Theorem 3, directly, without the aid of the Laurent expansion, by using the representation

$$f(z) = \frac{1}{2\pi i}\int_{\mathfrak{K}_1} \frac{f(\zeta)}{\zeta - z}\, d\zeta - \frac{1}{2\pi i}\int_{\mathfrak{K}_2} \frac{f(\zeta)}{\zeta - z}\, d\zeta,$$

where $\mathfrak{K}_1$ and $\mathfrak{K}_2$ are two suitable circles about z_0, and z lies in the annular region between them.

10. Prove the Casorati-Weierstrass theorem "$|f(z) - c| < \varepsilon$" (see K I, § 28, Theorem 5) by showing, with the help of Riemann's theorem demonstrated in the preceding problem, that the contrary assumption "$\left|\frac{1}{f(z) - c}\right| \leqq \frac{1}{\varepsilon}$" is untenable.

*11. Let the function $w = f(z)$ have an essential singularity at z_0. Then in every neighborhood of every complex number c there is another number a such that $f(z)$ has infinitely many a-points in a neighborhood of z_0.

§ 8. *The Residue Theorem, Zeros, and Poles*

1. With the aid of Rouché's theorem in K II, § 11, p. 111, prove the fundamental theorem of algebra by setting

$$\varphi(z) = a_0 + a_1 z + \cdots + a_{n-1} z^{n-1} \text{ and } f(z) = a_n z^n, \quad (a_n \neq 0),$$

and choosing for $\mathfrak{C}$ a circle with a sufficiently large radius.

2. The theorem, "If the roots of the entire *rational* function $g(z)$ are all real, then the roots of its derivative $g'(z)$ are also all real" (cf. § 1, Prob. 3) does *not* hold for entire transcendental functions. Prove this assertion by means of an example.

3. Determine the residues of

a) $\frac{1}{\sin z}$ at $z = k\pi$, $k = 0, \pm 1, \pm 2, \ldots$,

b) $\frac{z}{(z-1)(z-2)^2}$ at $z = +1$ and $z = +2$,

c) $\frac{a}{(z - z_1)^m (z - z_2)}$ at z_1 and $z_2 \neq z_1$, ($a \neq 0$, $m \geqq 1$ an integer),

d) $\tan z$ at $z = z_k = \frac{\pi}{2} + k\pi$, e) $e^{\frac{1}{z}}$ at $z = 0$,

f) $e^{\frac{1}{z-1}}$ at $z = +1$, g) $\dfrac{1}{1 - e^z}$ at $z = 2k\pi i$.

4. Let the function $f(z)$ have a zero of order α at z_0. What is the residue of

$$z\frac{f'(z)}{f(z)} \text{ and } \varphi(z)\frac{f'(z)}{f(z)}$$

at z_0, if $\varphi(z)$ denotes an arbitrary function which is regular at z_0? What is the answer if $f(z)$ has a pole of order β at z_0?

5. In connection with the preceding problem, determine the value and the meaning of the integrals

$$\frac{1}{2\pi i}\int_{\mathfrak{C}} z\frac{f'(z)}{f(z)}\,dz \text{ and } \frac{1}{2\pi i}\int_{\mathfrak{C}} \varphi(z)\frac{f'(z)}{f(z)}\,dz,$$

if the assumptions of Theorem 2 or of Theorem 3 in K I, § 33 are made concerning $f(z)$ and $\mathfrak{C}$.

6. Let $f(z)$ and $g(z)$ be regular at z_0. Suppose that $f(z_0) \neq 0$, whereas $g(z)$ has a zero of order 2 at z_0. What is the residue of $f(z)/g(z)$ at the point z_0? What is the answer if $g(z)$ has a zero of order 3 at z_0?

7. With the aid of the residue theorem, evaluate the integrals

a) $\displaystyle\int_{\mathfrak{C}} \frac{e^z}{z}\,dz$, where $\mathfrak{C}$ is the circle $|z| = 1$,

b) $\displaystyle\int_{\mathfrak{C}} \tan \pi z dz$, where $\mathfrak{C}$ is the circle $|z| = n$, $n = 1, 2, 3, \ldots$,

c) $\displaystyle\int_{\mathfrak{C}} \frac{f(z)}{(z - z_1)(z - z_2)\ldots(z - z_k)}$, where $\mathfrak{C}$ is the circle $|z| = R$,

under the assumption that the z_ν are distinct and all $|z_\nu| < R$ and that $f(z)$ is single-valued and regular in $|z| \leqq R$.

8. Let $R(x) = \dfrac{a_0 + a_1x + \cdots + a_mx^m}{b_0 + b_1x + \cdots + b_kx^k}$ be a rational function with real coefficients, whose denominator vanishes for no real value of x, the degree of the denominator exceeding that of the numerator by at least two units ($a_m \neq 0$, $b_k \neq 0$, $k \geqq m + 2$). Then the real integral

$$\int_{-\infty}^{+\infty} R(x)\,dx$$

converges and is equal to $2\pi i$ times the sum S of the residues of $R(z)$ at its poles in the upper half-plane.

9. In connection with I, § 7, Probs. 5 and 6, show that

$$I = \int_0^{+\infty} \frac{\sin x}{x}\,dx = \frac{\pi}{2}.$$

(Hint: The first and third of the integrals there together yield

$$2i \int_\rho^r \frac{\sin x}{x}\,dx.)$$

*10. By making use of the real integral $\int_0^\infty e^{-t^2}dt = \frac{1}{2}\int_0^\infty e^{-u}u^{-\frac{1}{2}}du = \frac{1}{2}\Gamma(\frac{1}{2}) = \frac{1}{2}\sqrt{\pi}$ (see K II, § 6, Ex. 3, (6)), evaluate Fresnel's integrals

$$\int_0^\infty \cos(t^2)\,dt = \int_0^\infty \sin(t^2)\,dt = \frac{1}{2}\sqrt{\frac{\pi}{2}}$$

by integrating the entire function e^{-z^2} along the closed path $\mathfrak{C}$ which extends rectilinearly from 0 to $+R$, thence along $|z| = R$ to the point $z_1 = Re^{i\frac{\pi}{4}}$, and thence rectilinearly back to 0. (Then let $R \to +\infty$.)

CHAPTER III

ENTIRE AND MEROMORPHIC FUNCTIONS

§ 9. *Infinite Products. Weierstrass's Factor-theorem*

1. Establish the convergence of, and evaluate, the following products of constant factors:

a) $\prod_{n=1}^{\infty}\left(1+\frac{1}{n(n+2)}\right)$; b) $\prod_{n=2}^{\infty}\left(1-\frac{2}{n(n+1)}\right)$;

c) $\prod_{n=2}^{\infty}\frac{n^3-1}{n^3+1}$; d) $\frac{2}{1}\cdot\frac{5}{4}\cdot\;\cdots\;\cdot\frac{n^2+1}{n^2}\cdot\;\cdots\;.$

2. Let the real numbers θ_n be wholly arbitrary except that they satisfy the condition $0<\theta_n<1$ for every $n=1, 2, 3, \ldots$. Then the series

$$\theta_1+[\theta_2(1-\theta_1)]+\cdots+[\theta_n(1-\theta_1)\cdots(1-\theta_{n-1})]+\cdots$$

is *invariably* convergent.

3. Prove Theorem 4 in K II, § 2 a) with, b) without, the use of the exponential function.

4. Let $\sum|a_n|^2$ be convergent. Then the two series $\sum a_n$ and $\sum \operatorname{Log}(1+a_n)$ are either both convergent or both divergent; here it is assumed that every $a_n \neq -1$, and Log denotes the principal value. In the first case, either both converge absolutely or both converge conditionally.

5. Determine the region of absolute convergence of the following products with variable terms:

a) $\prod_{n=1}^{\infty}(1-z^n)$; b) $\prod_{n=0}^{\infty}(1+z^{2^n})$;

c) $\prod_{n=0}^{\infty}(1+c_n z)$, if $\sum_{n=0}^{\infty}|c_n|$ converges;

d) $\prod\limits_{n=1}^{\infty}\left(1-\frac{1}{n^z}\right)$; e) $\prod\left(1-\frac{1}{p^z}\right)$,

where p runs through the succession of prime numbers.

6. For $|z|<1$,

a) $(1+z)(1+z^2)(1+z^4)(1+z^5)\cdots=\frac{z}{1-z}$;

b) $\frac{1}{(1-z)(1-z^3)(1-z^5)\cdots}=(1+z)(1+z^2)(1+z^3)\cdots$.

(Cf. K II, § 2, Theorem 6.)

7. Let the functions $f_n(z)$, $n=1, 2, \ldots$, be regular in the circle $|z|<r$, and let $\Sigma|f_n(z)|$ be uniformly convergent in every smaller circle $|z|\leqq\rho<r$, so that (according to K II, § 2, Theorem 7)

$$F(z)=\prod_{n=1}^{\infty}(1+f_n(z))=\sum_{\lambda=0}^{\infty}A_\lambda z^\lambda$$

is a regular function for $|z|<r$. Now set $f_n(z)=\sum\limits_{\lambda=0}^{\infty}a_\lambda^{(n)}z^\lambda$, and hence

$$F(z)=\sum_{\lambda=1}^{\infty}A_\lambda z^\lambda=\prod_{n=1}^{\infty}(1+a_0^{(n)}+a_1^{(n)}z+\cdots).$$

Then we assert that the series $\Sigma A_\lambda z^\lambda$ can be obtained by multiplying out the last product factor by factor. That is, if we set

$$(*)\qquad P_n(z)=\prod_{\nu=1}^{n}(1+f_\nu(z))=\sum_{\lambda=0}^{\infty}A_\lambda^{(n)}z^\lambda,$$

then

$$\lim_{n\to\infty}A_\lambda^{(n)}=A_\lambda$$

for every $\lambda=0, 1, 2, \ldots$.

8. On the basis of the theorem of the preceding problem, multiply out the product

$$\sin\pi z=\pi z\cdot\prod_{n=1}^{\infty}\left(1-\frac{z^2}{n^2}\right)$$

factor by factor, and compare the result with the power series for sin πz, — at least with respect to the coefficients of z^3 and z^5.

9. On the basis of the theorem in Prob. 7, expand the following products in power series:

a) $\prod_{n=1}^{\infty}(1 + c_n z) = \sum_{\nu=0}^{\infty} C_\nu z^\nu$, if $\Sigma|c_n|$ converges;

*b) $f(z) = \prod_{n=1}^{\infty}(1 + q^{2n-1}z) = \sum_{\nu=1}^{\infty} A_\nu z^\nu$, if $|q| < 1$;

Hint: We have $f(z) = (1 + qz)f(q^2 z)$, from which recursion formulas can be obtained by comparing coefficients.

c) $\prod_{n=1}^{\infty}(1 + c_n z)\left(1 + \frac{c_n}{z}\right) = \sum_{\nu=-\infty}^{+\infty} D_\nu z^\nu$, if $\Sigma|c_n|$ converges;

*d) $F(z) = \prod_{n=1}^{\infty}(1 + q^{2n-1}z)\left(1 + \frac{q^{2n-1}}{z}\right) = \sum_{\nu=-\infty}^{+\infty} B_\nu z^\nu$, if $|q| < 1$;

Hint: We have $F(z) = qzF(q^2 z)$, from which recursion formulas for the B_ν can again be obtained by comparing coefficients, enabling us to express $B_1, B_2, \ldots$ in terms of B_0. We get $B_\nu = q^{\nu^2} \cdot B_0$. From $q^{\nu^2} \cdot B_0 = A_\nu + A_1 A_{\nu+1} + \ldots$, which is obtained as in c), B_0 finally results as $B_0 = \lim_{\nu \to \infty} \frac{A_\nu}{q^{\nu^2}}$.

*e) $\prod_{n=1}^{\infty}(1 - z^n)$. Hint: In d), replace z by $-z^{1/2}$ and q by $z^{3/2}$, $|z| < 1$.

10. Derive the following from the sine-product:

a) $\sqrt{2} = \frac{2}{1} \cdot \frac{2}{3} \cdot \frac{6}{5} \cdot \frac{6}{7} \cdot \frac{10}{9} \cdot \frac{10}{11} \cdot \frac{14}{13} \cdot \cdots;$

b) $\sqrt{3} = 2 \cdot \frac{2}{3} \cdot \frac{4}{3} \cdot \frac{8}{9} \cdot \frac{10}{9} \cdot \frac{14}{15} \cdot \frac{16}{15} \cdot \cdots.$

11. Set up the product expansions for the following entire functions:

a) $e^z - 1$; b) $e^z - e^{z_0}$; \ c) $\cos \pi z$;

d) $\sin \pi z - \sin \pi z_0$; e) $\cos \pi z - \cos \pi z_0$.

*12. Prove the following transfer of Weierstrass's factor-theorem to the region of the unit circle:

Let $z_1, z_2, \ldots, z_n, \ldots$ be an arbitrary sequence of distinct points inside the unit circle, which have no limit point in the *interior* of this circle (but only on its circumference). Let $\alpha_1, \alpha_2, \ldots, \alpha_n, \ldots$ be a sequence of arbitrary positive integers. Then it is always possible to construct a function $f(z)$—and, indeed, in a form wholly analogous to the Weierstrass product—which is regular in the unit circle, and has zeros there, of the orders α_n, at precisely the respective points z_n (and at no others).

For further generalization to arbitrary regions instead of the unit circle, cf. § 11, Prob. 7.

13. With the aid of the theorem formulated in the preceding problem, construct functions which have the unit circle for a natural boundary but which are regular in its interior.

§ 10. *Entire Functions*

1. An entire function which assumes every value once, and only once, is an entire linear function.

2. The inverse of an entire function cannot be an entire function, except in the case of a linear function.

3. Prove Liouville's theorem that a bounded entire function $g(z)$ must be a constant, solely with the aid of the power-series expansion of entire functions (and hence, without the use of Cauchy's integral formula, which was employed in K I, § 28).

Hint: Consider the quotient $g_1(z) = \dfrac{g(z) - g(z_0)}{z - z_0}$ for large z, and apply Theorem 5 in K I, § 20.

4. Are there transcendental functions $g(z)$ for which $|g(z)| \to +\infty$ along *every* ray emanating from the origin?

5. Let the function $f(z)$ be regular at z_0. If the series $\sum_{n=0}^{\infty} f^{(n)}(z_0)$ converges, then $f(z)$ is an *entire* function, and the series $\sum_{n=0}^{\infty} f^{(n)}(z)$ converges for *every* z.

6. It is possible to find an entire function which assumes given values $w_1, w_2, \ldots, w_n, \ldots$ at given points $z_1, z_2, \ldots, z_n, \ldots$, provided that the latter have no finite limit point.

7. Let $\alpha \neq 0$ and β be arbitrarily given real numbers. Then there invariably exists an entire function $g(z) = \sum_{n=0}^{\infty} a_n z^n$ with *real rational* coefficients a_n, such that $g(\alpha) = \beta$.

8. The assertion made in the preceding problem remains valid if the word "real" is deleted in both places.

9. According to K I, § 20, Theorem 5, the maximum of the absolute value $|g(z)|$ of an entire function $g(z)$ in the circle $|z| \leqq r$ is attained at certain points of the boundary $|z| = r$. Denote this maximum by $M(r)$. Show that $M(r)$ is not only a monotonic function of the real variable r, but is also a continuous function of r.

10. Find the function $M(r)$, defined in the preceding problem, for the entire transcendental functions e^z, $\sin z$, $\cos z$, $\dfrac{\sin \sqrt{z}}{\sqrt{z}}$.

*11. Show by means of an example that the function $M(r)$ defined in the last problem but one does not have to be an analytic function of the real variable r.

§ 11. *Partial-fractions Series. Mittag-Leffler's Theorem*

1. Find the Mittag-Leffler partial-fractions-expansion for each of the following meromorphic functions:

a) $\tan z$, or, somewhat more conveniently, $\frac{\pi}{2} \tan \frac{\pi z}{2}$;

b) $\dfrac{1}{\sin z}$ or $\dfrac{\pi}{\sin \pi z}$; c) $\dfrac{\pi}{\cos \pi z}$;

d) $\dfrac{1}{e^z - 1}$; e) $\dfrac{\pi}{\cos \pi z - \sin \pi z}$.

2. Carry out in detail the derivation, sketched in K II, § 6, p. 47, of Weierstrass's theorem from Mittag-Leffler's.

3. Is Mittag-Leffler's theorem in the form given in K II, § 4, p. 37 still valid if the prescribed principal parts $h_\nu(z)$ are permitted to contain *infinitely* many negative powers of $(z - z_\nu)$? We are dealing then with functions which have only isolated singularities in the entire plane, and the question is whether at each of these points one can prescribe the descending part of the corresponding Laurent expansion, and if, or to what extent, such a function is thereby determined.

4. Can one—in connection with the preceding problem—prescribe at all or several or *one* of the points an initial section of the ascending

part, or even the entire ascending part, of the Laurent expansion?

*5. Transfer Mittag-Leffler's theorem to the unit circle (as Weierstrass's was transferred in § 9, Prob. 12); i.e., prove the following theorem: Let $z_0, z_1, \ldots, z_\nu, \ldots$ be any points for which $|z_\nu| < 1$ and $|z_\nu| \to 1$, and let the $h_\nu(z)$ be any principal parts (with a finite or an infinite number of negative powers). Then there exists a function $M_0(z)$ which is single-valued and regular for $|z| < 1$ except at the points z_ν, where it has isolated singularities with the $h_\nu(z)$ as the descending parts of the respective Laurent expansions.

*6. As a generalization of Mittag-Leffler's theorem and that of the preceding problem, prove the following theorem: Let $\mathfrak{M}$ be an infinite set of points which contains only isolated points and is therefore enumerable (proof?). Let $z_1, z_2, \ldots, z_\nu, \ldots$ be its points, and call $\mathfrak{M}'$ the set of limit points of $\mathfrak{M}$. With every z_ν associate a principal part $h_\nu(z)$. Then an infinite series can be set up, as in the case of Mittag-Leffler's theorem and the theorem of the preceding problem, which converges for every z which belongs neither to $\mathfrak{M}$ nor to $\mathfrak{M}'$. This series, moreover, represents in every region $\mathfrak{G}$ of the z-plane containing no point of $\mathfrak{M}'$ a single-valued analytic function which is regular everywhere in $\mathfrak{G}$ except at the points z_ν that lie in $\mathfrak{G}$. At these points the function possesses singularities with the principal parts $h_\nu(z)$.

*7. From the theorem in the preceding problem, derive an analogous extension of Weierstrass's factor-theorem, and formulate the theorem thus obtained.

8. Let $\mathfrak{M}$ and $\mathfrak{M}'$ have the same meaning as in Prob. 6, and let $w_1, w_2, \ldots$ be any complex numbers. Then one can always write down an expression which has a definite value at every point z which belongs neither to $\mathfrak{M}$ nor to $\mathfrak{M}'$, and which represents a regular analytic function in every region $\mathfrak{G}$ of the plane containing no point of $\mathfrak{M}'$, this function assuming at every point z_ν of $\mathfrak{G}$ the value w_ν.

9. In extension of the theorem stated in the preceding problem, prove the theorem (which in a certain sense settles this whole question) that the theorem formulated in Prob. 6 still holds if every one of the principal parts $h_\nu(z)$ prescribed there is allowed to contain a certain

number of terms with nonnegative powers, and is therefore of the form

$$H_\nu(z) = \sum_{k=-\infty}^{\beta_\nu} a_k^{(\nu)}(z - z_\nu)^k,$$

where the β_ν are given, nonnegative whole numbers.

§ 12. *Meromorphic Functions*

1. The sequence of functions

$$g_n(z) = \frac{n!\, n^z}{z(z+1)(z+2)\cdots(z+n)}, \quad n = 1, 2, \ldots,$$

converges *uniformly* (to $\Gamma(z)$, of course) in every bounded, closed region which contains none of the points 0, —1, —2, ...

2. Let z_1 and z_2 be two arbitrary distinct numbers different from 0, —1, —2, ... Under what conditions does

$$\lim_{n\to\infty} \frac{z_1(z_1+1)\cdots(z_1+n)}{z_2(z_2+1)\cdots(z_2+n)}$$

exist, and what is its value?

3. The so-called factorial series

$$F(z) = \sum_{n=1}^{\infty} \frac{n!\, a_n}{z(z+1)\cdots(z+n)}$$

converges at precisely the same points of the z-plane as the Dirichlet series

$$f(z) = \sum_{n=1}^{\infty} \frac{a_n}{n^z},$$

provided that the points 0, —1, —2, ... are left out of consideration.

4. The two series mentioned in the preceding problem are also uniformly convergent in precisely the same *regions* of the z-plane, provided that these are closed and contain none of the points 0, —1, —2, ...

5. The two series mentioned in the preceding problems are also *absolutely* convergent at precisely the same points of the z-plane (different from 0, —1, —2, ...).

6. Let $p_1, p_2, \ldots, p_n, \ldots$ be an entirely arbitrary sequence of positive integers. Then

$$\lim_{n\to\infty}\left\{\left(1+\frac{z}{n+1}\right)\left(1+\frac{z}{n+2}\right)\cdots\left(1+\frac{z}{n+p_n}\right)\right\}=1$$

if, and only if, $\frac{p_n}{n}\to 0$.

7. If $z = x + iy$ satisfies the condition $x \geqq -1$, $y \geqq 2$, then

$$\left|\frac{\Gamma'(z)}{\Gamma(z)}\right| < A \log |z|,$$

where A denotes a fixed positive constant (independent of z).

*8. Expand the Riemann ζ-function for a neighborhood of the point $+2$ in a power series

$$\zeta(z) = \sum_{n=0}^{\infty} (-1)^n b_n (z-2)^n,$$

and prove that $b_n \to 1$. What does this imply concerning the character of the point $+1$ for the ζ-function?

9. Show that for the coefficients b_n defined in the preceding problem the estimate

$$|b_n - 1| < \frac{A}{2^n}$$

holds, where A denotes a constant independent of n. What does this imply concerning the more precise nature of the point $+1$ for the ζ-function?

CHAPTER IV

PERIODIC FUNCTIONS

§ 13. *Simply Periodic Functions*

1. A (nonconstant) single-valued analytic function cannot have the periods 1 and $\sqrt{2}$.

2. A (nonconstant) rational function cannot be periodic.

3. If $f(z)$ has the primitive period 1 ($f(z)$ then is certainly not a constant), and if $f(z)$ exhibits a certain behavior as $z \to \infty$ in such a manner that $\mathfrak{J}(z) \to +\infty$ (or $\to -\infty$), then we say for brevity that $f(z)$ exhibits that behavior at the *upper* (or *lower*) end of the period strip. For the functions $f(z)$ of the class defined in K II, § 8, p. 70 we have then the following theorem:

If $f(z)$ is regular in the period strip, then it cannot be bounded at both ends.

4. If one of the functions $f(z)$ mentioned in the preceding problem is bounded at the upper (lower) end of the period strip, then $f(z)$ actually tends to a definite limit there. It is then meaningful to say that $f(z)$ assumes this value at that end of the strip. Under these conditions, how should one define the *order* to which $f(z)$ assumes the value a at an end of the strip?

5. If one of the functions $f(z)$ mentioned in Prob. 3 is not bounded at the upper (lower) end of the strip, then $f(z) \to \infty$ there. It is then meaningful to say that $f(z)$ has a pole at that end. Under these conditions, how should one define the order of such a pole?

6. Each of the functions $f(z)$ mentioned in Prob. 3 assumes every value, including the value ∞, if counted properly, equally often in the strip (including its ends).

7. A function having the primitive period $+1$ belongs to the class mentioned in Prob. 3 if, and only if, it has no singularities other than poles in the strip, including its ends.

§ 14. *Doubly Periodic Functions*

1. Let there be given a doubly periodic function's lattice of periods determined by the pair of primitive periods ω, ω'. Determine all pairs of primitive periods.

2. If $z_1, z_2, \ldots, z_k$ are the zeros, or, more generally, the c-points, of an elliptic function $f(z)$ in the period parallelogram, each taken as often as its multiplicity requires, and likewise if $\zeta_1, \zeta_2, \ldots, \zeta_k$ are the poles, then $\sum z_\kappa - \sum \zeta_\kappa$ is equal to a period of the function.

3. If $f(z)$ is an *odd* elliptic function, and $\tilde{\omega}$ is one of its periods, then $\frac{1}{2}\tilde{\omega}$ is either a zero or a pole of $f(z)$, and, in fact, necessarily one of *odd* order.

4. On which region in the w-plane does $w = e^{\frac{2\pi i}{\omega}z}$ map the fundamental parallelogram of a network of parallelograms in the z-plane, determined by (ω, ω'). On which region does it map the strip obtained from the fundamental parallelogram by means of the translations $(k'\omega')$?

5. According to K II, § 8, Theorem 1, every single-valued function $f(z)$ with the primitive period ω can be regarded as a single-valued function $\varphi(\zeta)$ of $\zeta = e^{\frac{2\pi i}{\omega}z}$. If $f(z)$ is a doubly periodic function with the pair of primitive periods (ω, ω'), what significance does this have for the function $\varphi(\zeta)$?

6. Is $e^{\wp(z)}$ an elliptic function?

*7. If ω is positive and real and ω' is positive and imaginary, then $\wp(z \mid \frac{1}{2}\omega, \frac{1}{2}\omega')$ is real on the boundary of the fundamental parallelogram and on its middle lines. On which region in the w-plane does $w = \wp(z \mid \frac{1}{2}\omega, \frac{1}{2}\omega')$ map the fundamental parallelogram in this case? Hint: Investigate the course of values of $\wp(z)$ on the boundary of the quarter of the fundamental parallelogram lying at 0.

*8. For the sums

$$s_n = \sum_{k, k'}{}' \frac{1}{(k\omega + k'\omega')^n}, \qquad n = 3, 4, 5, \ldots,$$

prove the relations

a) $s_{2m+1} = 0$ for $m = 1, 2, 3, \ldots$; b) $s_8 = \frac{3}{7}s_4^2$.

9. The function $w = \wp(z \mid \frac{1}{2}\omega, \frac{1}{2}\omega')$ satisfies (see K II, § 9, Theorem 9) the differential equation

$$(w')^2 = 4w^3 - g_2 w - g_3,$$

with $g_2 = 60 s_4$, $g_3 = 140 s_6$, if s_n has the meaning given in Prob. 8. Show that the roots of the cubic equation

$$4w^3 - g_2 w - g_3 = 0$$

have the values $\wp(\frac{1}{2}\omega)$, $\wp(\frac{1}{2}\omega')$, $\wp(\frac{1}{2}(\omega + \omega'))$ and are distinct.

10. In connection with K II, § 9, Theorem 10, and Prob. 2, show that every elliptic function $f(z)$ can be represented as a "σ-quotient", i.e., with the aid of the σ-function belonging to the same period-parallelogram, in the form

$$c\,\frac{\sigma(z - z_1)\,\sigma(z - z_2)\cdots\sigma(z - z_k)}{\sigma(z - \zeta_1)\,\sigma(z - \zeta_2)\cdots\sigma(z - \zeta_k)}.$$

CHAPTER V

ANALYTIC CONTINUATION

§ 15. *Behavior of Power Series on the Boundary of the Circle of Convergence*

1. Let the power series $h(z) = \sum_{n=0}^{\infty} b_n z^n$ have radius $r = 1$, let $b_n > 0$ and $\sum b_n$ diverge, and let it be related to the power series $f(z) = \sum_{n=0}^{\infty} a_n z^n$ so that $\frac{a_n}{b_n} \to g$. Then $\sum a_n z^n$ has a radius $r' \geqq 1$, and if $z \to +1$ radially, then also

$$\lim \frac{f(z)}{h(z)} = g.$$

*2. Is the theorem formulated in the preceding problem still valid, or under what additional conditions is it valid, if the variable approaches $+1$ but remains within an "angular region" $z_1 z_2 1$ as in I, § 11, Prob. 9?

3. With the aid of the theorem formulated in Probs. 1 and 2, prove Abel's limit theorem (I, § 11, Prob. 10) once more by setting $h(z) = \frac{1}{1-z}$ and $f(z) = \frac{1}{1-z} \sum a_n z^n = \sum s_n z^n$, $s_n = a_0 + a_1 + \cdots + a_n$.

4. With the aid of the theorem formulated in Probs. 1 and 2, prove the following extension of Abel's limit theorem: If the coefficients a_n of the power series $F(z) = \sum_{n=0}^{\infty} a_n z^n$ have the property that, if we set $a_0 + a_1 + \cdots + a_n = s_n$,

$$\frac{s_0 + s_1 + \cdots + s_n}{n+1} \to s,$$

then also $F(z) \to s$ if z tends to $+1$ within an angular region $z_1 z_2 1$ (cf. Prob. 2).

5. If z tends to $+1$ radially (or, as in Prob. 2, within an angular region $z_1 z_2 1$), then

a) $(1 - z + z^4 - z^9 + z^{16} - + \cdots) \to \frac{1}{2}$,

b) $\sqrt{1-z}\,(1 + z + z^4 + z^9 + \cdots) \to \frac{1}{2}\sqrt{\pi}$,

c) $\dfrac{1}{\log \dfrac{1}{1-z}}\,(z + z^p + z^{p^2} + z^{p^3} + \cdots) \to \dfrac{1}{\log p}$ (p an integer $\geqq 2$),

d) $(1-z)^{p+1}\,(z + 2^p z^2 + 3^p z^3 + \cdots) \to \Gamma(p+1)$
($p > -1$, arbitrary).

6. Let the power series $f(z) = \sum_{n=0}^{\infty} a_n z^n$ have radius 1 and converge at $z = +1$: $a_0 + a_1 + \cdots + a_n = s_n \to s$. If this convergence is so strong that $\sqrt{n}\cdot(s_n - s) \to 0$, then (cf. Prob. 3) we still have $f(z) \to s$ if z tends to $+1$ but remains within the ellipse

$$x^2 + \frac{y^2}{\alpha^2} = 1, \qquad (0 < \alpha < 1).$$

7. The two series

$$f_1(z) = z + \frac{z^3}{3} - \frac{z^4}{2} + \frac{z^5}{5} + \frac{z^7}{7} - \frac{z^8}{4} + + - \cdots$$

$$f_2(z) = z + \frac{z^3}{3} - \frac{z^2}{2} + \frac{z^5}{5} + \frac{z^7}{7} - \frac{z^4}{4} + + - \cdots$$

are obviously convergent for $|z| < 1$ and for $z = +1$. The first is an ordinary power series, whereas the second is a rearrangement of a power series. What values do the series have at $+1$, and what limiting values do the functions represented by these series for $|z| < 1$ have as $z \to +1$ radially?

8. The function $(1-z)\cdot\sin\left(\operatorname{Log}\dfrac{1}{1-z}\right)$, where Log denotes the principal value, can be expanded in a power series for $|z| < 1$. Show that this series converges absolutely for $|z| = 1$.

§ 16. *Analytic Continuation of Power Series*

1. Show that the function defined by the power series

$$\mathfrak{P}(z) = \sum_{n=1}^{\infty} \frac{z^n}{n}$$

is multiple-valued, by continuing the series analytically, according to K I, § 24, in the negative sense about the point $+1$ until one returns to 0 – without assuming any properties of the function defined by $\mathfrak{P}(z)$ to be known. (Cf. K I, p. 104, Fig. 8.) Hint: Choose the points $z_1, z_2, \ldots$ on the circle $|z - 1| = 1$ at the vertices of an inscribed regular p-gon ($p > 6$), one of whose vertices lies at the origin. Then $z_\nu = 1 - z^{-\frac{2\nu\pi i}{p}}$, $\nu = 1, 2, \ldots, p$, and the expansion about $z_p = 0$ obtained after p steps differs from $\mathfrak{P}(z)$ only in that an additive constant has appeared. One finds immediately that this constant is equal to $2\pi i$, by letting $p \to \infty$.

2. Suppose that the power series $f(z) = \sum a_n z^n$ is known to be such that the function $f(z)$ which it represents has only *one* singular point z_0 on the boundary of the circle of convergence, and that this singular point is a simple pole. Show then that

$$\frac{a_n}{a_{n+1}} \to z_0, \text{ and hence } \left|\frac{a_n}{a_{n+1}}\right| \to r,$$

the radius of the power series.

3. Let the power series $\mathfrak{P}(z) = \sum_{n=0}^{\infty} a_n z^{n+1}$ have radius 1. Let $z = \frac{\zeta}{1-\zeta}$, expand every term $a_n z^{n+1}$ in powers of ζ, and arrange the resulting double series according to powers of ζ to form the power series $\mathfrak{P}_1(\zeta)$. How do the coefficients of $\mathfrak{P}_1(\zeta)$ read, and, accordingly, i.e., by the Cauchy-Hadamard theorem, what is its radius? (Cf. § 2, Prob. 3b.) How can the radius of $\mathfrak{P}_1(\zeta)$ be read off, on the other hand, from the analytic properties of $f(z)$? Consequently, what is the *least* possible value and the *greatest* possible value of the radius?

4. Under the conditions of the preceding problem, what is a

necessary and sufficient condition for the function $f(z)$ represented by $\mathfrak{P}(z)$ in $|z|<1$ to be regular at $+1$?

5. Show with the aid of the considerations in the two preceding problems that the functions represented by the series

$$\sum_{n=0}^{\infty}(-1)^n z^{n+1} \quad \text{and} \quad \sum_{n=0}^{\infty}(-1)^n \frac{z^{n+1}}{n+1}$$

are regular at $+1$.

6. With the aid of the criterion established in Prob. 4, prove once more the theorem in I, § 11, Prob. 3.

7. Extend the theorem in I, § 11, Prob. 3, which was reconsidered in the preceding problem, to obtain the following theorem: Let the power series $f(z) = \Sigma a_n z^n$ have radius 1, and suppose that the same is true of the power series $\varphi(z) = \Sigma \alpha_n z^n$, where $\alpha_n = \Re(a_n)$, and that invariably $\alpha_n \geqq 0$. Then $+1$ is a singular point of the function $f(z)$ represented by $\Sigma a_n z^n$.

§ 17. *Analytic Continuation of Arbitrarily Given Functions*

1. Can the real function $F(x) = \sqrt{x^2} = |x|$, defined for $-\infty < x < +\infty$, be continued into the complex domain?

2. Can the real function defined for $-1 < x < +1$ by

$$f(x) = \begin{cases} e^{-\frac{1}{x^2}} & \text{for } x \neq 0 \\ 0 & \text{for } x = 0 \end{cases}$$

be continued into the complex domain? (Note that $f(x)$ possesses continuous derivatives of all orders at every point of the interval of definition! Cf. § 7, Prob. 5.)

3. Let $g_0, g_1, \ldots, g_n, \ldots$ be a sequence of positive integers, each of which is $\geqq 2$. Then the series

$$\sum_{n=0}^{\infty} \frac{z^{g_0 g_1 \cdots g_n}}{1 - z^{g_0 g_1 \cdots g_n}}$$

represents a function $f(z)$ which is regular for $|z| < 1$ but which cannot be continued beyond the unit circle. (Cf. I, § 11, Probs. 4 and 5.)

4. Let the function $f(z)$ be regular in a neighborhood of the origin, and let

$$(*) \qquad f(2z) = 2f(z) \cdot f'(z)$$

there. Then $f(z)$ can be continued over the entire plane, and is therefore an entire function. (Cf. $f(z) = \sin z$.) What is the sole essential property of the right-hand side of (*) for the proof of this assertion?

5. Let the function $f(z)$ be regular for $\mathfrak{R}(z) > 0$, satisfy the relation

$$(\dagger) \qquad f(z+1) = zf(z)$$

there, and let $f(1) \neq 0$. Then $f(z)$ is a meromorphic function whose only poles lie at $0, -1, -2, \ldots$ and are simple. What residues has $f(z)$ there? (Cf. the function $\Gamma(z)$.)

6. Prove the following extension of Riemann's theorem (K I, § 32, Theorem 3): Let $f(z)$ be continuous in $\mathfrak{G}$ and, save perhaps at the points of a broken line $\mathfrak{w}$ lying in $\mathfrak{G}$, also differentiable. Then $f(z)$ is necessarily regular at these points too. (The endpoints of $\mathfrak{w}$ may also lie on the boundary of $\mathfrak{G}$.)

7. Let the two regions $\mathfrak{G}_1$ and $\mathfrak{G}_2$ abut along the broken line $\mathfrak{w}$ (i.e., let the points of $\mathfrak{w}$ be boundary points of both regions) but have no point in common. Let $f_1(z)$ be regular in $\mathfrak{G}_1$ and $f_2(z)$ be regular in $\mathfrak{G}_2$. If (cf. I, § 5, Prob. 8) f_1 and f_2 take on the same boundary values along $\mathfrak{w}$, then they are analytic continuations of one another.

8. The functional element $f(z)$ represented by the power series $\sum \frac{z^n}{n}$ can be continued in the sense of the monodromy theorem (K I, § 25) along every path in the region $1 < |z-2| < 3$. It does not, however, generate a single-valued function in this region. Does this contradict the monodromy theorem?

9. Show the following by means of an example: If, under the general assumptions of the monodromy theorem (K I, § 25), $f(z)$ can be continued to every other point z_1 of $\mathfrak{G}$, not along *every* path in $\mathfrak{G}$ but only along *suitable* paths, then the function which is thus generated in $\mathfrak{G}$ is not necessarily single-valued and regular.

10. Extend the monodromy theorem (K I, § 25) by showing that if $f(z)$ exhibits no singularities other than poles in the course of

arbitrary continuation within $\mathfrak{G}$, then $f(z)$ generates a function which is *single-valued* in $\mathfrak{G}$ and is regular there except for poles.

11. Extend the monodromy theorem (K I, § 25) finally by showing that if $f(z)$ can be continued arbitrarily close to every point ζ in $\mathfrak{G}$, and if the continuation obtained is single-valued and regular in $0 < |z - \zeta| < \delta$ for a $\delta = \delta(\zeta) > 0$, then $f(z)$ generates a function which is single-valued in $\mathfrak{G}$ and regular there except for isolated points.

CHAPTER VI

MULTIPLE-VALUED FUNCTIONS AND RIEMANN SURFACES

§ 18. *Multiple-valued Functions in General*

1. Can a function in a region be everywhere regular and yet not single-valued? (Answer the question for simply and multiply connected regions.)

2. Is it possible for a multiple-valued function to have the same value at two superposed points of its Riemann surface? Can it have the same value at infinitely many such points? Can it have the same value at all points of a neighborhood of two such points?

3. What kind of function (single-valued or multiple-valued) is defined by each of the following formulas:

a) $\sqrt{e^z}$; b) $\sqrt{\cos z}$; c) $\sqrt{1-\sin^2 z}$;

d) $\sqrt{\wp(z)-\wp(\frac{1}{2}\omega)}$; e) $\sqrt{\wp(z)}$; f) $\log e^z$;

g) $\log \sin z$; h) $\dfrac{\sin\sqrt{z}}{\sqrt{z}}$?

4. What values can the integral $\int_0^{z_0} \frac{dz}{\sqrt{1-z^2}}$ have if the path lies in the simple z-plane, avoids the points ± 1, and leads in an arbitrary manner from the point 0 to a definite point z_0 ($\neq \pm 1$)? Here the initial value of the radical (i.e., its value at the initial point 0 of the path) is to be taken as $+1$. — Do these integrals still have a meaning for $z_0 = \pm 1$ or for $z_0 = \infty$? (Hint: Use § 1, Prob. 6 to reduce the paths to simpler ones.)

5. What values can the integrals

$$\int_1^{z_0} \frac{dz}{\sqrt[p]{z}} \quad \text{and} \quad \int_1^{z_0} \log z \, dz$$

have for an arbitrary path extending from the point $+1$ of a definite sheet to the point z_0 of a certain sheet and avoiding the origin (p an integer $\geqq 1$)?

6. (Cf. I, § 11, Prob. 4d.) Let

$$f(z) = z + \sum_{n=1}^{\infty} \frac{z^{2^n+2}}{(2^n+1)(2^n+2)}$$

and

$$F_1(z) = \sqrt{f(z)}, \quad F_2(z) = \log f(z).$$

Construct the Riemann surfaces of these functions. (Hint: Prove first that, in $|z|<1$, if $z \neq 0$, then also $f(z) \neq 0$.)

7. Let $f(z)$ denote the same function as in the preceding problem. Construct the Riemann surfaces of the two functions

$$G_1(z) = \log f\left(\frac{z}{2}\right) + \log f\left(\frac{1}{2z}\right)$$

and

$$G_2(z) = \log f\left(\frac{z}{2}\right) - \log f\left(\frac{1}{2z}\right).$$

§ 19. *Multiple-valued Functions; in Particular, Algebraic Functions*

1. Construct the Riemann surfaces for the following functions:

a) $\sqrt[3]{z-a}$; b) $\sqrt[3]{(z-a)^2}$; c) $\sqrt[3]{\frac{z-a}{z-b}}$;

d) $\sqrt[3]{(z-a)(z-b)}$; e) $\sqrt[3]{(z-a)(z-b)(z-c)}$;

f) $\sqrt[3]{(z-a_1)(z-a_2)\cdots(z-a_k)}$, $k>3$;

g) $\sqrt[3]{\frac{(z-a)}{(z-b)^2}} + \sqrt{z-c}$; h) $\sqrt[n]{(z-a)(z-b)(z-c)}$, $n>3$.

2. Construct the Riemann surfaces for the following functions:

a) $\log(z-a)$; b) $\log(z-a)(z-b)$;

c) $\log\frac{z-a}{z-b}$; d) $\log(1+z^2)$;

e) $\operatorname{arc\,tan} z$.

3. Picture the Riemann surfaces of the functions

$$z^i \text{ and } z^a, \quad (a \text{ an arbitrary complex number}).$$

*4. Study in all details the construction of the Riemann surfaces (critical points, connection of the sheets there, behavior of the function at these points, distribution of the domain of values, etc.) of the algebraic functions w of z defined by the following equations:

a) $w^3 - z - 1 = 0$; b) $w + \frac{1}{w} - z = 0$;

c) $w^n + \frac{1}{w^n} - z = 0$; d) $w^3 - 3w - z = 0$;

e) $w^3 + 3w - z = 0$.

CHAPTER VII

CONFORMAL MAPPING

§ 20. *Concept and General Theory*

1. Let $f(z)$ be regular for $|z-z_0| < r$, and let $|z_1-z_0| < r$, $z_1 \neq z_0$, $f(z_0) = w_0$, $f(z_1) = w_1$. Through what angle does the segment $\overline{w_0w_1}$ appear to be rotated, and in what ratio does it appear to be stretched, relative to the segment $\overline{z_0z_1}$?

2. Let $f(z)$ have a pole at z_0. When can a mapping of a neighborhood of z_0 (including this point itself) on a neighborhood of the point $w_0 = \infty$ by means of $w = f(z)$ be called conformal? What is the situation if also $z_0 = \infty$? What is the answer if $z_0 = \infty$ but w_0 is finite?

3. Let the function $f(z)$ be regular along the simple closed path $\mathfrak{C}$ and in the region $\mathfrak{G}$ enclosed by $\mathfrak{C}$. Let $w = f(z)$ map $\mathfrak{C}$ on the simple closed path $\mathfrak{C}'$ in a one-to-one manner (so that if z traverses the path $\mathfrak{C}$ once in the positive sense, the image point w also describes the path $\mathfrak{C}'$ once in a definite sense). Then the closed region $\mathfrak{G}$ is mapped in a one-to-one manner on the closed region $\mathfrak{G}'$ enclosed by $\mathfrak{C}'$, and the mapping of the interior regions on each other is conformal.

4. Is the theorem stated in the preceding problem still true if the curve $\mathfrak{C}'$ passes through ∞? How should the previously assumed regularity of $f(z)$ along $\mathfrak{C}$ be formulated now? Is the theorem still true, finally, if $\mathfrak{C}$ likewise passes through ∞?

5. Let the bounded path $\mathfrak{k}$, including its endpoints, lie in a region of regularity of the function $w = f(z)$, and let $\mathfrak{k}'$ be the image of $\mathfrak{k}$ in the w-plane. Is $\mathfrak{k}'$ also a path? What is the length of $\mathfrak{k}'$?

6. Let $f(z)$ be regular for $|z| < r$, and let $0 < \rho < r$. What is the area $J(\rho)$ of the image $\mathfrak{K}'$ of the disk $\mathfrak{K}$ with radius ρ and center 0? Is the formula which is to be derived still valid if the inverse of the mapping of $\mathfrak{K}$ on $\mathfrak{K}'$ is not *single-valued*, i.e., if $f'(z)$ has zeros in $\mathfrak{K}$, so that $\mathfrak{K}'$ is multiple-sheeted?

7. Under the conditions of the preceding problem, what is the limit of the ratio of the areas of $\mathfrak{R}'$ and $\mathfrak{R}$ as $\rho \to 0$? Is the answer still valid if $f'(0) = 0$? (Areal ratio of magnification.)

8. Let the function $f(z)$ be regular in $|z| \leqq r$. Determine

a) the length of the image of the circumference $|z| = r$;

b) the area of the image of the disk $|z| \leqq r$;

c) the direction of the tangent to the image of $|z| = r$ at a point $w = f(z)$ for which $f'(z) \neq 0$;

d) the curvature of the image of $|z| = r$ at a point $w = f(z)$ for which $f'(z) \neq 0$.

9. What changes must be made in the answers to questions a) to d) of the preceding problem if $f(z)$ is regular for $|z| \geqq r$ including $z = \infty$?

*10. Let $w = \varphi(\zeta)$, regular for $|\zeta| < 1$, map the interior of the unit circle in a one-to-one manner on the region $\mathfrak{G}$ of the w-plane. Let $\varphi(0) = w_0$, and let $\overline{\mathfrak{G}}_\rho$ be the image of the circle $|\zeta| \geqq \rho < 1$. Suppose that $f(z)$ is a function which is regular in $|z| < 1$ and is such that the values w which it assumes there lie in $\mathfrak{G}$, and, in particular, is such that $f(0) = w_0$. Then, for $|z| \leqq \rho$, the point $f(z) = w$ lies in $\overline{\mathfrak{G}}_\rho$, and actually in the interior of this region unless $f(z) = \varphi(e^{i\alpha} z)$, α fixed. [Hint: If $\zeta = \Phi(w)$ is the inverse of the function $w = \varphi(\zeta)$, apply Schwarz's lemma, B, p. 29, to $\Phi(f(z))$.]

11. Let $f(z)$ be regular in $|z| < 1$, and let $\Re f(z) > 0$ there. Set $f(0) = w_0 = u_0 + iv_0$. Then in $|z| \leqq \rho < 1$ we have actually

$$u_0 \frac{1-\rho}{1+\rho} \leqq \Re f(z) \leqq u_0 \frac{1+\rho}{1-\rho}.$$

[Hint: In the preceding problem, take the region $\mathfrak{G}$ to be the right half-plane $\Re(w) > 0$.]

12. What can be asserted about $\Im f(z)$ and $|f(z)|$ under the conditions of the preceding problem?

13. The function $f(z) = a_0 + a_1 z + a_2 z^2 + \cdots$, regular in the unit circle, is certainly *simple* (*schlicht*) there, if

$$2|a_2| + 3|a_3| + \cdots \leqq |a_1|.$$

Cf. § 18, Prob. 6.

*14. The limit theorem for simple mappings (B, § 17) asserts the following: Let every function of the sequence $f_1(z), f_2(z), \ldots, f_n(z), \ldots$ be single-valued and regular in the region $\mathfrak{G}$ and effect a simple mapping of this region. In every closed subregion $\overline{\mathfrak{G}'}$ of $\mathfrak{G}$, let the $f_n(z)$ converge uniformly to a limit function $f(z)$ which is not identically constant. Then the function $f(z)$ (which is regular in the entire region $\mathfrak{G}$) also effects a simple mapping of $\mathfrak{G}$.

Prove this theorem with the aid of the theorem on zeros formulated in § 5, Prob. 4, without further recourse to Cauchy's integral formula.

15. The inequality

$$\frac{|z|}{(1+|z|)^2} \leqq |f(z)| \leqq \frac{|z|}{(1-|z|)^2},$$

proved in B, p. 106, can be sharpened by replacing the denominators on the left and on the right by $(1+|z|^2)$ and $(1-|z|^2)$, respectively, if we make the additional assumption concerning $f(z)$ that it is an odd function $(f(z) = -f(-z))$. [Hint: Show that $(f(z))^2$ is a simple function of $z^2 = \zeta$ and apply the above inequality to this function.]

16. If a region $\mathfrak{G}$ of the z-plane is to be mapped in a one-to-one, conformal manner on a circle in the w-plane with 0 as center in such a way that a certain point a of $\mathfrak{G}$ goes over into $w = 0$ and that the linear ratio of magnification at this point equals $+1$, then the radius $r = r(\mathfrak{G}; a)$ of the image circle in the w-plane is hereby uniquely determined.

§ 21. *Specific Mapping Problems*

1. In connection with I, § 12, Prob. 20, discuss in detail the following closure problem: Let $\mathfrak{k}_1, \mathfrak{k}_2, \mathfrak{k}_3, \ldots$ be circles which lie in the region between $\mathfrak{K}_1$ and $\mathfrak{K}_2$, i.e., in the intersection of the exterior regions of the two circles, where the exterior of each of the circles denotes that part of the plane in which the other circle lies. The circle $\mathfrak{k}_1$ is chosen arbitrarily there, except that $\mathfrak{K}_1$ and $\mathfrak{K}_2$ are tangent to $\mathfrak{k}_1$ and lie in one and the same region of the two determined by $\mathfrak{k}_1$, which we again regard as the exterior of $\mathfrak{k}_1$. Let the succeeding circles $\mathfrak{k}_\nu$, $\nu = 2, 3, \ldots$, be required to be such that $\mathfrak{k}_\nu$ is tangent to the circles $\mathfrak{K}_1$, $\mathfrak{K}_2$, and

$\mathfrak{k}_{\nu-1}$, and that these three circles lie in one and the same region determined by $\mathfrak{k}_\nu$ (the "exterior" of $\mathfrak{k}_\nu$). Then for $\mathfrak{k}_2$ there are still two possibilities; for $\nu \geqq 3$, $\mathfrak{k}_\nu$ is uniquely determined by the requirement of being different from $\mathfrak{k}_{\nu-2}$. Show that one is hereby either always (i.e., no matter how first $\mathfrak{k}_1$ and then $\mathfrak{k}_2$ be chosen) or never led to infinitely many different circles $\mathfrak{k}_\nu$, and that in the second case the chain of circles $\mathfrak{k}_\nu$ always "closes" after the same number of terms.

2. Let the circle $\mathfrak{K}_1$ mentioned in the preceding problem have center $z_1 = \frac{17}{21} - \frac{i}{7}$ and radius $r_1 = \frac{8}{15}$, and let $\mathfrak{K}_2$ have center $z_2 = \frac{69}{25} + \frac{33}{25}i$ and radius $r_2 = \frac{24}{7}$. Determine whether in this case the chain referred to in the preceding problem closes or not,. and, if it does close, after how many terms.

3. If the same (nonconstant) linear transformation is repeatedly applied to z, i.e., if we set

$$\frac{az+b}{cz+d} = z', \quad \frac{az'+b}{cz'+d} = z'', \quad \frac{az''+b}{cz''+d} = z''', \quad \ldots,$$

then every $z^{(n)}$ is a linear function of z:

$$(L_n) \qquad z^{(n)} = \frac{a_n z + b_n}{c_n z + d_n}.$$

Show that, for every n,

$$\frac{b_n}{b} = \frac{c_n}{c} = \frac{a_n - d_n}{a - d},$$

and characterize the mapping effected by (L_n).

4. Show that the stereographic projection considered in I, § 12, Probs. 6–9 and Elem., § 14 effects a *conformal* mapping of the plane and the sphere on one another.

5. The so-called Mercator projection associates with the point P of the sphere (the earth) having geographical coordinates λ (longitude) and β (latitude) the point P' of an xy-plane with the (Cartesian) coordinates $x = R\lambda$, $y = R \log \tan\left(\frac{\pi}{4} + \frac{\beta}{2}\right)$; $-\pi < \lambda \leqq +\pi$, $-\frac{\pi}{2} < \beta < +\frac{\pi}{2}$. Show that this Mercator projection effects a conformal mapping of the sphere on the plane. (R is the radius of the sphere.)

6. Map the plane cut along the negative axis of reals from $-\frac{1}{4}$ to $-\infty$ on the interior of the unit circle in such a manner that the origin and the directions emanating from it remain fixed.

7. Investigate the mapping of the z-plane effected by the function $w = f(z)$ which is defined by means of

$$aw^2 + bw + c = z, \qquad (a \neq 0).$$

8. Map the simply connected region containing all points of the plane (including ∞) with the exception of the real points z in $-2 \leqq z \leqq +2$, on the interior of the unit circle.

9. Map the exterior region of the cardioid

$$\begin{cases} x = 2a\,(1 - \cos t)\cos t, \\ y = 2a\,(1 - \cos t)\sin t, \end{cases} \qquad t = 0 \ldots 2\pi, \qquad a > 0,$$

conformally on the interior of the unit circle. [Hint: Reflect the cardioid with respect to the unit circle.]

10. Map the rectangle whose vertices lie at the points $\pm 2 \pm i$, on the unit circle. (Cf. § 14, Prob. 7.)

*11. Map the equilateral triangle whose vertices lie at the points $0, 1, \frac{1}{2} + \frac{i}{2}\sqrt{3}$, on the upper half-plane. [Hint: Carry out considerations for the integral $\int_0^z t^{-\frac{2}{3}} \cdot (1-t)^{-\frac{2}{3}} dt$ analogous to those in B, § 14.]

*12. The integrals

$$w = f_1(z) = \int_0^z \frac{t\,dt}{t^3 - 1} \quad \text{and} \quad f_2(z) = \int_0^z \frac{dt}{t^3 - 1}$$

are infinitely multiple-valued. Pick out a single-valued branch by cutting the plane in a suitable manner, and investigate the mapping of the cut plane which is effected by the branch. Cut the plane so that the image is a simple (single-sheeted) domain.

13. In connection with § 20, Prob. 16, determine the radius $r = r(\mathfrak{G}; a)$ if $\mathfrak{G}$ is the region lying between the hyperbola $xy = 1\,(x > 0,\ y > 0)$ and the positive coordinate axes, and if $a = \alpha + \dfrac{i}{2\alpha}\ (\alpha > 0)$.

14. In connection with § 20, Prob. 16, at every point a of $\mathfrak{G}$ imagine a perpendicular of length $r(\mathfrak{G}; a)$ to be erected. What kind of surface do the endpoints of this perpendicular form if $\mathfrak{G}$ is

a) the circle $|z - z_0| < R$,

b) the half-plane $\mathfrak{J}(z) > 0$,

c) the strip $0 < \mathfrak{J}(z) < \pi$,

d) the region mentioned in the preceding problem?

PART II - SOLUTIONS

CHAPTER I

ADDITIONAL PROBLEMS FOR I, CHS. 1–5

§ 1. *Fundamental Concepts*

1. The proposition asserts that every (closed) half-plane $\mathfrak{H}$ containing $z_1, z_2, \ldots, z_k$ also contains ζ. If $\mathfrak{H}$ is the half-plane $\Re(z) \geqq 0$, then $\Re(z_\nu) \geqq 0$, $\nu = 1, 2, \ldots, k$, and hence also $\Re(\zeta) \geqq 0$, so that the proposition is true. Every other half-plane $\mathfrak{H}'$ can be transformed into the right half-plane by means of a translation and a rotation. This means that a and α (α real) can be chosen so that $\Re e^{i\alpha}(z_\nu - a) \geqq 0$, $\nu = 1, 2, \ldots, k$. By the preceding case, since $\sum \alpha_\nu = 1$, we have also $\Re e^{i\alpha}(\zeta - \alpha) \geqq 0$, so that ζ itself must lie in $\mathfrak{H}'$.—The extension to infinite sequences is obvious.

2. If we put $1/(\zeta - z_\nu) = z_\nu'$, then $\alpha_1 z_1' + \cdots + \alpha_k z_k' = 0$. This means that $\Re(z_\nu')$, and hence also $\Re(1/z_\nu')$, for $\nu = 1, 2, \ldots, k$, either has positive and negative values or else is always equal to 0. If the equation above is multiplied by $e^{i\alpha}$ (α real), we see that every straight line through 0 either separates the points $1/z_\nu'$ or else contains them all. If the translation (ζ) is now performed, we see that every straight line through ζ either contains all points z_ν or else *both* sides of the line contain at least one point. Consequently, ζ cannot lie outside the polygon in question.

3. If $G'(\zeta) = 0$, then also

$$\frac{1}{n} \cdot \frac{G'(\zeta)}{G(\zeta)} = \frac{1}{n}\left(\frac{\alpha_1}{\zeta - z_1} + \frac{\alpha_2}{\zeta - z_2} + \cdots + \frac{\alpha_k}{\zeta - z_k}\right) = 0,$$

where we have set $G(z) = c(z - z_1)^{\alpha_1}\ (z - z_2)^{\alpha_2} \cdots\ (z - z_k)^{\alpha_k}$, $\alpha_1 + \alpha_2 + \cdots + \alpha_k = n$. The proof is now completed by referring to Problem 2.

4. a) Two distinct linear transformations, each of which sends z_ν to ∞, differ merely by a similarity transformation, which does not

alter a_ν, the number of vertices. We may therefore employ particular transformations. Pass to the Riemann sphere by stereographic projection, and bring z_ν to ∞ by means of a rotation. The polygon π_ν now is obviously projected from ∞ by a solid angle of a_ν edges. Since the rotation does not alter configurational relations, it is seen that $a_1 + a_2 + \cdots + a_n$ is twice the number of edges of the convex polyhedron of n vertices determined by the points z_ν (on the sphere). If we denote the number of its edges by e and the number of its faces by f, then $2e = 3f$, because all faces are triangles (since no four of the z_ν lie on a circle). According to Euler's theorem on polyhedra, $n + f = e + 2$. These two equalities yield the assertion $2k = a_1 + \cdots + a_n = 6(n - 2)$.

b) If every $a_\nu \geqq 6$, a) is contradicted.

c) and d) Let x_p denote the *number* of p-gons among the π_ν. Then, in addition to $x_3 + x_4 + \cdots + x_{n-1} = n$, we have, according to a),

$$3x_3 + 4x_4 + \cdots + (n - 1)x_{n-1} = 6(n - 2).$$

Multiplying the first equation by 6 and subtracting the result from the second, it follows that $3x_3 + 2x_4 + x_5 \geqq 12$, from which both assertions can be read off.

5. Of the $\binom{n}{3}$ planes which are determined by triples of vertices of the polyhedron discussed in connection with the preceding problem, the $f = 2(n - 2)$ faces, and only these, have the property that all the remaining $n - 3$ vertices lie on only one of the two spherical caps determined by the plane. If this is transferred to the Gaussian *plane*, the assertion follows.

6. See K I, p. 15, Lemma 1.

§ 2. *Infinite Sequences and Series*

1. Since $\lim_{n\to\infty} (b_0 - b_n) = \sum_{n=0}^{\infty} (b_n - b_{n+1})$, and hence $\lim b_n$ exists, $\{b_n\}$ is bounded. If invariably $|b_n| < K$, say, then

$$|b_n^q - b_{n+1}^q| = |b_n - b_{n+1}| \cdot |b_n^{q-1} + b_n^{q-2}\, b_{n+1} + \cdots + b_{n+1}^{q-1}|$$
$$\leqq q \cdot K^{q-1} \cdot |b_n - b_{n+1}|;$$

hence, $\sum |b_n^q - b_{n+1}^q|$ converges. The proof is now completed according to I, § 3, Prob. 12.

2. a) and b). Let $\sum a_n$ converge absolutely and $\sum b_n$ at least conditionally. Let the partial sums of the three series be A_n, B_n, C_n, and let their sums be A, B, and, if it exists, C. Then $C_n = a_0 B_n + a_1 B_{n-1} + \cdots + a_n B_0$. If we now set $B_n = B + \beta_n$, so that $\beta_n \to 0$, then

$$C_n = A_n \cdot B + (a_0\beta_n + a_1\beta_{n-1} + \cdots + a_n\beta_0),$$

which, according to I, § 3, Prob. 9, tends to $A \cdot B$.

c) He have

$$\frac{C_0 + C_1 + \cdots + C_n}{n+1} = \frac{A_0 B_n + A_1' B_{n-1} + \cdots + A_n B_0}{n+1}.$$

According to I, § 3, Prob. 2, the left-hand side $\to C$, and by I, § 3, Prob. 7b, the right-hand side $\to AB$. Consequently, $AB = C$.

3. Since $a_0 + a_1 + \cdots + a_n = s_n \to s$, we have also, according to I, § 3, Prob. 2, 8, respectively:

a) $$s_n' = \frac{s_0 + s_1 + \cdots + s_n}{n+1} \to s$$

and

b) $$s_n'' = \frac{1}{2^{n+1}} \left\{ \binom{n+1}{1} s_0 + \binom{n+1}{2} s_1 + \cdots + \binom{n+1}{n+1} s_n \right\} \to s.$$

But these are precisely the assertions.

[Note that $\frac{1}{2^{\nu+1}} \binom{\nu}{\nu} + \frac{1}{2^{\nu+2}} \binom{\nu+1}{\nu} + \ldots + \frac{1}{2^{n+1}} \binom{n}{\nu} = \frac{1}{2^{n+1}} \left\{ \binom{n+1}{\nu+1} + \cdots + \binom{n+1}{n+1} \right\}$.]

4. For every real $\alpha \neq 0$ the sequence $\{n^{-i\alpha}\}$ obviously diverges. Therefore the series $\sum_1^\infty a_n$, with $a_1 = 1$, $a_n = n^{-i\alpha} - (n-1)^{-i\alpha}$ for $n \geqq 2$, also diverges.

Now

$$a_n = n^{-i\alpha} \left[1 - \left(1 - \frac{1}{n} \right)^{-i\alpha} \right] = n^{-i\alpha} \left[-\frac{i\alpha}{n} + \frac{r_n}{n^2} \right],$$

where $\{r_n\}$ denotes a bounded sequence $\left[\text{because } r_n \to -\left(\frac{-i\alpha}{2}\right)\right]$.

Consequently,

$$a_n = -i\alpha \frac{1}{n^{1+i\alpha}} + \frac{r_n}{n^{2+i\alpha}},$$

from which the assertion can be read off.

5. The solution of the preceding problem shows that the series $\Sigma n^{-(1+i\alpha)}$, for $\alpha \neq 0$, has bounded partial sums. According to I, § 3, Prob. 13a, the new series is certainly convergent.

6. We have

$$1 - \binom{z}{1} + \binom{z}{2} = 1 - z + \frac{z(z-1)}{1 \cdot 2} = (1-z)\left(1 - \frac{z}{2}\right)$$

and in general

$$1 - \binom{z}{1} + \binom{z}{2} - + \cdots + (-1)^n \binom{z}{n} = (1-z)\left(1 - \frac{z}{2}\right) \cdots \left(1 - \frac{z}{n}\right).$$

Now we can set $1 - \frac{z}{k} = e^{-\frac{z}{k} + \frac{A_k}{k^2}}$ with bounded A_k (for fixed z).

Consequently, except for a convergent factor with a nonzero limit, the sequence of partial sums of our series behaves like the sequence of numbers

$$e^{-z\left(1 + \frac{1}{2} + \cdots + \frac{1}{n}\right)},$$

and therefore converges for a $z \neq 0$, and in fact to the limit 0, if, and only if, $\Re(z) > 0$.

7. Set $\sum_{n=1}^{\infty} a_n \frac{z^n}{1-z^n} = f(z)$ and $\sum_{n=1}^{\infty} a_n z^n = g(z)$. Then, since

$$\frac{z^n}{1-z^n} = z^n + z^{2n} + \cdots + z^{kn} + \cdots,$$

we have $f(z) = \sum_{k=1}^{\infty} g(z^k)$. If we put in turn

$$a_n = (-1)^{n-1}, = n, = \frac{(-1)^{n-1}}{n}, = \alpha^n,$$

we obtain the identities a) to d). The series in question converge uniformly for $|z| \leqq \rho < 1$ according to I, § 9, Prob. 1f; and Weierstrass's double-series theorem (K I, p. 83) allows us to perform the necessary rearrangements.

8. We may assume that $\beta = 1$, $r > 1$ (proof?). If we set $b_n/b_{n+1} = 1 + \varepsilon_n$, then $\varepsilon_n \to 0$. Further,

$$\left|\frac{c_{n+1}}{b_{n+1}} - f(1)\right| \leqq |a_1 \varepsilon_n| + |a_2| \{(1 + |\varepsilon_n|)(1 + |\varepsilon_{n-1}|) - 1\} + \cdots$$
$$+ |a_{n+1}| \{(1 + |\varepsilon_n|) \cdots (1 + |\varepsilon_0|) - 1\} + \sum_{\nu=n+2}^{\infty} |a_\nu|.$$

If r_1 is now chosen in $1 < r_1 < r$ and held fixed, and m is determined so that $1 + |\varepsilon_\nu| < r_1$ for $\nu > m$, then for $n > p > m$ the above expression is

$$\leqq \sum_{\nu=0}^{p} |a_{\nu+1}| \{(1 + |\varepsilon_n|) \cdots (1 + |\varepsilon_{n-\nu}|) - 1\}$$
$$+ (1 + |\varepsilon_0|) \cdots (1 + |\varepsilon_m|) \sum_{\nu=p+2}^{\infty} |a_\nu| r_1^\nu.$$

Now let $\varepsilon > 0$ be given.

Then one can first determine p so that the second term here is $< \frac{1}{2}\varepsilon$, and then determine $n > p$ so that the first term is also $< \frac{1}{2}\varepsilon$.

9. The only solutions of the equation $\zeta = \alpha\zeta + \dfrac{\beta}{\zeta}$ are ± 1, so that only these two points come into question as possible limits.

First, if $\Re(z_0) = 0$, then also $\Re(z_1) = 0$ or $z_1 = \infty$. The points $z_0, z_1, z_2, \ldots$ thus remain on the axis of imaginaries or are ∞. Therefore they cannot $\to \pm 1$, so the sequence must diverge.

Since the situation is the same for $\Re(z_0) < 0$ as for $\Re(z_0) > 0$, it suffices to assume that $\Re(z_0) > 0$ for the rest of the proof. Let $\mathfrak{S}$ be the family of circles whose centers lie on the positive axis of reals and which intersect the unit circle orthogonally. If z lies on the boundary of the circle $\mathfrak{K}$ of $\mathfrak{S}$, then $1/z$ lies there too, and consequently $z' = \alpha z + (\beta/z)$ lies in its interior. Hence, if z_0 lies in $\mathfrak{K}_0$, then all the rest of the points of the sequence $\{z_n\}$ lie there too. We now show that a point ζ which lies in the right half-plane and is different from $+1$ cannot be a limit point of the sequence $\{z_n\}$. For if $z \to \zeta$, then $z' = \alpha z + (\beta/z)$ tends to $\zeta' = \alpha\zeta + (\beta/\zeta)$. One can therefore describe such a small circle $\mathfrak{k}$ about ζ, that $\mathfrak{k}$ does not contain $+1$, and that for all z of $\mathfrak{k}$ the corresponding z' lies in that circle $\mathfrak{K}$ of the family $\mathfrak{S}$ which is tangent to $\mathfrak{k}$ exteriorly. Thus if z_p lies in $\mathfrak{K}$, so does z_{p+1}, and hence also $z_{p+2}, z_{p+3}, \ldots$. Therefore ζ cannot be a limit point of $\{z_n\}$. Consequently, $z_n \to 1$.

10. It suffices to show that every closed half-plane $\mathfrak{H}$ which contains all ζ also contains all ζ'. To this end we consider also the half-planes $\mathfrak{H}_1, \mathfrak{H}_2, \mathfrak{H}_3$ which project beyond $\mathfrak{H}$ by parallel strips of width ε, 2ε, 3ε ($\varepsilon > 0$). Choose p so that for $\nu > p$ the z_ν lie in $\mathfrak{H}_1$, and for $n > p$ set

$$z_n'' = a_{np+1}\, z_{p+1} + \cdots + a_{nn}\, z_n,$$

so that $z_n' - z_n'' \to 0$. According to § 1, Prob. 1, $z_n''/(a_{np+1} + \cdots + a_{nn})$ also lies in $\mathfrak{H}_1$ for $n > p$.

Since the denominator $\to +1$, z_n'' lies in $\mathfrak{H}_2$ for $n > p_1 > p$, and hence z_n' lies in $\mathfrak{H}_3$ for $n > N$ because $z_n' - z_n'' \to 0$. Thus no ζ' can lie outside $\mathfrak{H}_3$. Consequently, no ζ' can lie outside $\mathfrak{H}$ either, since such a point, for sufficiently small ε, would also lie outside $\mathfrak{H}_3$.

§ 3. *Functions of a Complex Variable*

1. a) Yes. An example is $f(z) = |z|^2$. We have $\left|\dfrac{f(z) - f(0)}{z - 0}\right| = |z| \to 0$. It is obvious that $|z|^2$ is not differentiable at the points $z \neq 0$.

b) Yes. An example is $f(z) = (\mathfrak{J}(z))^2$. For if $z = x + iy$ and if $z_0 = x_0$ is real, then, as $z \to z_0$,

$$\left|\frac{f(z) - f(z_0)}{z - z_0}\right| = \frac{y^2}{|(x - x_0) + iy|} \leq |y| \to 0.$$

At nonreal points $f(z)$ is obviously not differentiable.

2. Yes. An example is $f(z) = f(x + iy)$

$$= \begin{cases} 0 \text{ if } y \text{ is irrational}, = 1 \text{ if } y = 0, \\ \dfrac{1}{q^3} \text{ if } y = \dfrac{p}{q} \; (p,\, q \text{ relatively prime integers, } p \neq 0,\; q > 0). \end{cases}$$

At the points with rational ordinates, $f(z)$ is discontinuous, and hence certainly not differentiable. If, however, $z_0 = x_0 + iy_0$ has an irrational ordinate, then

$$\left|\frac{f(z) - f(z_0)}{z - z_0}\right| \leqq \frac{1}{q^3} \cdot \frac{1}{\left|\frac{p}{q} - y_0\right|} \quad \text{or} = 0$$

according as $\mathfrak{J}(z) = y$ equals p/q or is irrational. Now if y_0 is a root of a quadratic equation with integral coefficients, then, as is well known[1]), for every $\frac{p}{q}$ invariably

$$\left|\frac{p}{q} - y_0\right| \leqq \frac{c}{q^2}, \qquad (c > 0 \text{ depending only on } y_0).$$

Hence the difference quotient $\to 0$ as $z \to z_0$. The points z_0, however, with a y_0 of the kind considered, lie everywhere dense in the plane (proof?).

3. No, unless the function reduces to a constant. For if, in $f(z) = u + iv$, v, say, is constant in $\mathfrak{G}_1$, then, according to the Cauchy-Riemann differential equations, u is also constant there. Consequently $f(z)$ itself is constant. By considering $\log f(z)$, the other questions are reduced to the one just considered.

4. Let M be a bound for $|f'(z)|$ in $|z| < 1$, and let $\varepsilon > 0$. For two interior points z' and z'' of the unit circle at a distance less than $\varepsilon/2M$ from a point ζ with $|\zeta| \leqq 1$ we have

$$|f(z'') - f(z')| = \left|\int_{z'}^{z''} f'(z)\,dz\right| \leqq M \cdot |z'' - z'| < \varepsilon.$$

Thus the assumptions of I, § 5, Prob. 8 are fulfilled. The rest is answered by I, § 5, Probs. 8 and 9.

5. Set $\sqrt{x + iy} = u + iv$. Then

$$\left.\begin{aligned} u^2 - v^2 &= x \\ 2uv &= y \end{aligned}\right\} \text{ and hence } u^2 + v^2 = \sqrt{x^2 + y^2} \text{ (to be taken } \geqq 0).$$

[1]) If $\varphi(y) = ky^2 + my + n$ (k, m, n real integers) has the irrational (real) roots y_0 and y_1, then

$$\left|\varphi\left(\frac{p}{q}\right)\right| = \left|k\left(\frac{p}{q} - y_0\right)\left(\frac{p}{q} - y_1\right)\right| \geqq \frac{1}{q^2}.$$

From this it follows (proof?) that for a suitable $c > 0$ depending only on k, m, n,

$$\left|\frac{p}{q} - y_0\right| \geqq \frac{c}{q^2}.$$

Therefore $u = \pm\sqrt{\frac{1}{2}(\sqrt{x^2+y^2}+x)}$ and v equals that one of the two values $\pm\sqrt{\frac{1}{2}(\sqrt{x^2+y^2}-x)}$ for which uv has the sign of y (because $2uv = y$). For $y = 0$ the formula gives $\pm\sqrt{x}$ or $\pm i\sqrt{|x|}$ according as $x > 0$ or < 0; and for $x = y = 0$, the unique value 0.

6. If the exponent is not a real integer, e^z is *uniquely* defined by the sum of the series, whereas for $a \neq e$ the power a^z is *multiple-valued.* E.g., $e^{\frac{1}{2}}$ is single-valued, $(-e)^{\frac{1}{2}}$ is two-valued.

7. a) If we set $z - 1 = \rho(\cos\varphi + i\sin\varphi)$ with $\frac{\pi}{2} < |\varphi| \leqq \pi$, then

$$f(z) = e^{\left(\frac{1}{1-z}\right)^2} = e^{\frac{1}{\rho^2}(\cos 2\varphi - i\sin 2\varphi)}.$$

Hence, if $\frac{3}{4}\pi < |\varphi| \leqq \pi$, then $f(z) \to \infty$; if $\frac{\pi}{2} < |\varphi| < \frac{3}{4}\pi$, then $f(z) \to 0$.

For $\varphi = \pm\frac{3}{4}\pi$ we have indefinite divergence.

b) For $z = \cos t + i\sin t$ we have

$$\left(\frac{1}{1-z}\right)^2 = \frac{1}{4}\left[1 - \cot^2\frac{t}{2} + 2i\cot\frac{t}{2}\right]; t \neq 2k\pi.$$

Hence, $f(z) \to 0$.

8. a) If $z = re^{i\varphi}$, $-\pi < \varphi \leqq +\pi$, then

$$z^a = r^a \cdot e^{ia\varphi + 2ka\pi i},\ k = 0, \pm 1, \pm 2, \ldots .$$

This is a uniquely determined value if, and only if, ka is an integer for $k = 0, \pm 1, \ldots$, which is obviously the case only for integral a.

b) For $z = re^{i\varphi}$, $-\pi < \varphi \leqq +\pi$, the principal value is

$$z^i = e^{-\varphi + i\ln r};\ |z^i| = e^{-\varphi} < e^{\pi}.$$

9. The branch in question is

$$(1-z)^i = e^{i\log(1-z)},\ |(1-z)^i| = e^{-\mathrm{am}(1-z)},$$

where the principal value is taken for the logarithm and the amplitude. For the latter in $|z| < 1$ we have, however,

$$-\frac{\pi}{2} < \mathrm{am}(1-z) < +\frac{\pi}{2}.$$

10. If $z = re^{i\varphi}$ lies in the right half-plane, then

$$\operatorname{Log} z = \ln r + i\varphi \text{ with } -\tfrac{\pi}{2} \leqq \varphi \leqq +\tfrac{\pi}{2}$$

and

$$\operatorname{Log}^2 z = (2\varphi \ln r)i - \vartheta, \text{ with } \vartheta \text{ real.}$$

Consequently,

$$\exp(-i \operatorname{Log}^2 z) = r^{2\varphi} \cdot e^{i\vartheta},$$

$$z^3 \exp(-i \operatorname{Log}^2 z) = r^{3+2\varphi} \cdot e^{i(\vartheta+3\varphi)}.$$

Therefore the function $\to 0$ if the constant φ lies in $-\tfrac{3}{2} < \varphi \leqq \tfrac{\pi}{2}$; it becomes infinite if $-\tfrac{\pi}{2} \leqq \varphi < -\tfrac{3}{2}$; and it is indefinitely divergent, but remains bounded, if z tends to 0 along the ray $\varphi = -\tfrac{3}{2}$. (Draw a figure.)

§ 4. *Integral Theorems*

1. a) Since a rectifiable path can have one or even infinitely many corners, the answer to the first two questions is yes. b) The same for the third! For there exist (in the real domain) monotonic, continuous functions, $y = f(x)$, which are nondifferentiable for values of x which are everywhere dense. The most familiar example of such a function is the one given by H. A. Schwarz,

$$f(x) = \sum_{n=0}^{\infty} \frac{1}{4^n} \varphi(2^n x),$$

where $\varphi(x)$ is the function $[x] + \sqrt{x - [x]}$ and $[x]$ denotes the greatest integer $\leqq x$ (H. A. Schwarz, *Gesammelte mathematische Abhandlungen*, vol. II, p. 269). Its graph in the xy-plane yields a rectifiable path (because $f(x)$ is continuous and monotonic) which lacks a tangent at an everywhere dense set of points. — This cannot occur at *all* points, however. The proof of this requires deeper theorems in the theory of real functions: 1. $x = x(t)$, $y = y(t)$ yields a rectifiable path if, and only if, the functions $x(t)$ and $y(t)$ are continuous and of bounded variation. 2. A continuous function of bounded variation can be written as the sum of two continuous, monotonic functions. 3. A continuous, monotonic function $f(x)$ is differentiable for all x except, perhaps, at the points of a set of (Lebesgue) measure 0.

2. It is sufficient to show that every half-plane $\mathfrak{H}$ which contains $\mathfrak{k}'$ also contains the quotient μ of the two integrals. This quotient, however, is the limit of

$$\frac{\sum_{\nu=1}^{n} (t_\nu - t_{\nu-1})\, \rho(\zeta_\nu) \cdot f(\zeta_\nu)}{\sum_{\nu=1}^{n} (t_\nu - t_{\nu-1}) \rho(\zeta_\nu)} = \sum_{\nu=1}^{n} \alpha_\nu \cdot f(\zeta_\nu).$$

Since the values $f(\zeta_\nu)$ lie in $\mathfrak{H}$, and $\alpha_\nu > 0$ and $\sum_{\nu=1}^{n} \alpha_\nu = 1$, the value of this expression also lies in $\mathfrak{H}$, according to § 1, Prob. 1, — and, in fact, for every n. The same is true, therefore, for the limit μ.

3. The existence proof in K I, § 9 is also valid here with almost no change. Only in the proof of Lemma 1, the differences $(b - a)$, $(a_1 - a), \ldots, (b - a_{p-1})$ must from the first be enclosed in absolute-value signs. — Check carefully all the details.

4. We have

$$\left| \sum_{\nu=1}^{n} (z_\nu - z_{\nu-1}) f(\zeta_\nu) \right| \leqq \sum_{\nu=1}^{n} |z_\nu - z_{\nu-1}| \cdot |f(\zeta_\nu)|,$$

from which the assertion is obtained by passing to the limit.

5. Let us set $\bar{U}(t) = U(x(t), y(t))$, and analogously $\bar{V}(t)$. Then, using the notation in K I, § 10,

$$\sum_{\nu=1}^{n} |z_\nu - z_{\nu-1}| \cdot F(\zeta_\nu) =$$

$$\sum_{\nu=1}^{n} (t_\nu - t_{\nu-1}) \sqrt{(x'(\tau_\nu))^2 + (y'(\tau_\nu))^2}\, (\bar{U}(\tau_\nu) + i\bar{V}(\tau_\nu)).$$

This tends to

$$\int_\alpha^\beta (\bar{U}(t) + i\bar{V}(t)) \sqrt{(x'(t))^2 + (y'(t))^2}\, dt$$

$$= \int_\alpha^\beta \bar{U}(t) \sqrt{x'^2 + y'^2}\, dt + i \int_\alpha^\beta \bar{V}(t) \sqrt{x'^2 + y'^2}\, dt$$

$$= \int_\mathfrak{k} U ds + i \int_\mathfrak{k} V ds = \int_\mathfrak{k} (U + iV)\, ds.$$

6. Let $\mathfrak{K}$ be a circle about ζ with radius $\rho_1 < \rho$. If m is so large that both z_n and z_n' lie in $\mathfrak{K}$ for $n > m$, then

$$f(z_n) = \frac{1}{2\pi i}\int_{\mathfrak{K}} \frac{f(z)}{z - z_n}\, dz, \quad f(z_n') = \frac{1}{2\pi i}\int_{\mathfrak{K}} \frac{f(z)}{z - z_n'}\, dz,$$

$$\frac{f(z_n') - f(z_n)}{z_n' - z_n} - f'(\zeta) = \frac{1}{2\pi i}\int_{\mathfrak{K}} f(z)\left[\frac{1}{(z - z_n)\,(z - z_n')} - \frac{1}{(z - \zeta)^2}\right] dz.$$

From this it follows, exactly as in the proof of the integral formulas for the derivatives in K I, § 16, that the left-hand side $\to 0$ as $z_n \to \zeta$ and at the same time $z_n' \to \zeta$.

7. Since $F(z, t)$ is regular in $\mathfrak{G}$ for fixed t on $\mathfrak{k}$,

$$\frac{1}{2\pi i}\int_{\mathfrak{K}} \frac{F(\zeta, t)}{\zeta - z}\, d\zeta = F(z, t).$$

The iterated integral given in the hint to the solution is thus equal to $f(z)$. Likewise,

$$F_z(z, t) = \frac{1}{2\pi i}\int_{\mathfrak{K}} \frac{F(\zeta, t)}{(\zeta - z)^2}\, d\zeta.$$

From this representation one can read off, in particular, the fact that $F_z(z, t)$ is a continuous function of t along $\mathfrak{k}$ for fixed z in $\mathfrak{G}$. Hence, if z_0 lies in $\mathfrak{G}$, $\mathfrak{K}$ denotes a circle lying in $\mathfrak{G}$ about z_0, and $z \neq z_0$ denotes another point inside $\mathfrak{K}$, then

$$\frac{f(z) - f(z_0)}{z - z_0} - \int_{a,\,\mathfrak{k}}^{b} F_z(z_0, t)\, dt$$

$$= \frac{1}{2\pi i}\int_{a,\,\mathfrak{k}}^{b} \left\{ \int_{\mathfrak{K}} F(\zeta, t)\left[\frac{1}{(\zeta - z)\,(\zeta - z_0)} - \frac{1}{(\zeta - z_0)^2}\right] d\zeta \right\}.$$

From this it follows, exactly as in K I, § 16, that, as $z \to z_0$, the expression $\to 0$. Consequently, $f'(z_0)$ exists and has the asserted value.

§ 5. *Expansions in Series*

1. By assumption, about every point z of $\mathfrak{G}$ one can describe a circle $\mathfrak{K}_z$ such that the series converges uniformly in $\mathfrak{K}_z$ (or in that

part of $\mathfrak{K}_z$ which belongs to $\mathfrak{G}$). According to the Heine-Borel theorem (K I, § 3, Theorem 3), a finite number of these circles, say $\mathfrak{K}_1$, $\mathfrak{K}_2$, ..., $\mathfrak{K}_p$, are sufficient to cover $\mathfrak{G}$. Consequently, if $\varepsilon > 0$ is given, one can choose n_ν ($\nu = 1, 2, \ldots, p$) in such a manner that for all $n > n_\nu$, all $k \geqq 1$, and all z in $\mathfrak{K}_\nu$,

$$|f_{n+1}(z) + f_{n+2}(z) + \cdots + f_{n+k}(z)| < \varepsilon.$$

If N is greater than each of the numbers $n_1, \ldots, n_p$, this inequality is obviously satisfied for all z in $\mathfrak{G}$ provided that $n > N$ and $k \geqq 1$. This, however, is the assertion.

2. If z is given in $\mathfrak{G}$, describe a circle $\mathfrak{K}$ about z, whose circumference likewise lies wholly within $\mathfrak{G}$. Then, by assumption, $\sum_{n=0}^{\infty} f_n(\zeta)$ is uniformly convergent in the closed disk $\mathfrak{K}$. On the circumference of $\mathfrak{K}$, $\sum_{n=0}^{\infty} \frac{f_n(\zeta)}{\zeta - z}$ is then also uniformly convergent and equal to $\frac{f(\zeta)}{\zeta - z}$.

According to K I, § 19, Theorem 2, then,

$$\frac{1}{2\pi i} \int_{\mathfrak{K}} \frac{f(\zeta)}{\zeta - z}\, d\zeta = \sum_{n=0}^{\infty} \frac{1}{2\pi i} \int_{\mathfrak{K}} \frac{f_n(\zeta)}{\zeta - z}\, d\zeta = \sum_{n=0}^{\infty} f_n(z) = f(z).$$

By K I, § 16, however, the integral on the left defines a function which is regular inside $\mathfrak{K}$, and hence, in particular, at z. Consequently, $f(z)$ is regular in $\mathfrak{G}$. Furthermore, according to this § 16,

$$f^{(p)}(z) = \frac{p!}{2\pi i} \int_{\mathfrak{K}} \frac{f(\zeta)}{(\zeta - z)^{p+1}}\, d\zeta = \sum_{n=0}^{\infty} \frac{p!}{2\pi i} \int_{\mathfrak{K}} \frac{f_n(\zeta)}{(\zeta - z)^{p+1}}\, d\zeta,$$

this last because the series employed converges uniformly with respect to ζ on the circumference of $\mathfrak{K}$. Therefore $\sum_{n=0}^{\infty} f_n^{(p)}(z)$ converges and equals $f^{(p)}(z)$. That the last series actually converges uniformly in every $\mathfrak{G}'$ now follows as in K I, p. 76.

3. Let $\mathfrak{G}'$ be given, and let $\mathfrak{C}$ be chosen as in K I, p. 76. Then if $\varepsilon > 0$ is given, n_0 can be determined so that, for $n > n_0$, $k \geqq 1$, and ζ on $\mathfrak{C}$, we have

(*) $$|f_{n+1}(\zeta)| + |f_{n+2}(\zeta)| + \cdots + |f_{n+k}(\zeta)| < \varepsilon.$$

For all z in $\mathfrak{G}'$, then,

$$|f^{(p)}_{n+1}(z)| + \cdots + |f^{(p)}_{n+k}(z)| \leqq \sum_{\nu=n+1}^{n+k} \frac{p!}{2\pi} \int_{\mathfrak{C}} \frac{|f_\nu(\zeta)|}{|\zeta - z|^{p+1}} |d\zeta|,$$

where the integrals on the right-hand side are understood in the sense of § 4, Prob. 3. Denote the (certainly positive) greatest lower bound of the distance $|\zeta - z|$ between a point of $\mathfrak{C}$ and a point of $\mathfrak{G}'$ by ρ, and the length of $\mathfrak{C}$ by l. Then, because of (*), the last sum is

$$< \frac{p!}{2\pi} \cdot \frac{\varepsilon}{\rho^{p+1}} \cdot l,$$

from which the assertion follows.

4. Let ζ lie within $\mathfrak{G}$. Then one can describe such a small circle $\mathfrak{K}$, with radius ρ, about ζ, that its circumference also lies wholly within $\mathfrak{G}$, and that inside and on the boundary of $\mathfrak{K}$, $F(z) \neq 0$, except, perhaps, at ζ itself. According to K I, § 33, Theorem 2, we have then

(a) $$\frac{1}{2\pi i} \int_{\mathfrak{K}} \frac{F'(z)}{F(z)} dz — \;= 0 \text{ or } = \alpha$$

according as $F(\zeta) \neq 0$ or vanishes to the order α. This integral differs from the integral

(b) $$\frac{1}{2\pi i} \int_{\mathfrak{K}} \frac{s_n'(z)}{s_n(z)} dz$$

by arbitrary little provided merely that n is sufficiently large. For if we denote by $\mu > 0$ the greatest lower bound of $|F(z)|$ on the boundary of K, and choose a positive $\varepsilon < \frac{1}{2}\mu$, then an n_0 can be determined so that, for $n > n_0$ and all z on $\mathfrak{K}$, invariably $|s_n(z) - F(z)| < \varepsilon$ and at the same time, according to K I, § 19, also $|s_n'(z) - F'(z)| < \varepsilon$. Then obviously $|s_n(z)| > \frac{1}{2}\mu$ along $\mathfrak{K}$, so that the integral (b) exists. Furthermore, the absolute value of the difference of (a) and (b) is

$$\leqq \frac{1}{2\pi} \cdot 2\pi\rho \cdot \max \left| \frac{F'(z)}{F(z)} - \frac{s_n'(z)}{s_n(z)} \right|,$$

the maximum taken on the boundary of $\mathfrak{K}$; and hence, is

$$\leqq \rho \frac{2M\varepsilon}{\frac{1}{2}\mu\cdot\mu},$$

if M denotes an upper bound of $|F(z)|$ and $|F'(z)|$ on $\mathfrak{K}$. Since ε may be taken arbitrarily small, and the difference of (a) and (b) is an integer, this integer must equal 0. Thus, for all sufficiently large n, $s_n(z)$ has inside $\mathfrak{K}$ exactly the same number of zeros as $F(z)$. This proves the assertions of the theorem, including a more precise result concerning the *number* of zeros of $s_n(z)$ in a neighborhood of ζ.

5. Concerning the first part of the problem, cf. § 3, Prob. 9. Further,

$$nb_n = n\,\frac{i(i+1)\,(i+2)\cdots(i+n-1)}{1\cdot 2\cdot 3\cdots n},$$

$$|nb_n| = \left|1+\frac{i}{1}\right|\cdot\left|1+\frac{i}{2}\right|\cdots\left|1+\frac{i}{n-1}\right|$$

$$= \left[\left(1+\frac{1}{1^2}\right)\left(1+\frac{1}{2^2}\right)\cdots\left(1+\frac{1}{(n-1)^2}\right)\right]^{\frac{1}{2}}.$$

This, according to K II, § 3, Ex. 1, tends to

$$\left(\prod_{n=1}^{\infty}\left(1+\frac{1}{n^2}\right)\right)^{\frac{1}{2}} = \left(\frac{\sin \pi i}{\pi i}\right)^{\frac{1}{2}} = \left(\frac{e^{\pi}-e^{-\pi}}{2\pi}\right)^{\frac{1}{2}}.$$

6. We have

$$(1+z)^{\frac{1}{z}} = e^{\frac{1}{z}\operatorname{Log}(1+z)} = e^{1-\frac{z}{2}+\frac{z^2}{3}-+\cdots} = e\cdot e^{-\frac{z}{2}}\cdot e^{+\frac{z^2}{3}}\cdots$$

$$= e\cdot\left[1-\frac{z}{2}+\frac{z^2}{2!\,2^2}-\frac{z^3}{3!\,2^3}+\frac{z^4}{4!\,2^4}-\frac{z^5}{5!\,2^5}+-\cdots\right]$$

$$\cdot\left[1+\frac{z^2}{3}+\frac{z^4}{2!\,3^2}+\cdots\right]\left[1-\frac{z^3}{4}+\cdots\right]$$

$$\cdot\left[1+\frac{z^4}{5}+\cdots\right]\left[1-\frac{z^5}{6}+\cdots\right]\cdots$$

$$= e\left[1-\tfrac{1}{2}z+\tfrac{11}{24}z^2-\tfrac{7}{16}z^3+\tfrac{2147}{5760}z^4-\tfrac{959}{2304}z^5+\cdots\right].$$

7. We have

$$f(z) = 1 + \sum_{\nu=1}^{\infty} \frac{\alpha^\nu}{\nu!} \left(\frac{z}{1-z}\right)^\nu = 1 + \sum_{\nu=1}^{\infty} \sum_{\lambda=0}^{\infty} \frac{\alpha^\nu}{\nu!} \binom{\nu+\lambda-1}{\lambda} z^{\nu+\lambda}$$

$$= 1 + \sum_{n=1}^{\infty} \left\{ \sum_{k=1}^{n} \binom{n-1}{k-1} \frac{\alpha^k}{k!} \right\} z^n.$$

Hence, $c_0 = 1$, and, for $n \geqq 1$,

$$c_n = \sum_{k=1}^{n} \binom{n-1}{k-1} \frac{\alpha^k}{k!} = \gamma_1(n) + \gamma_2(n) + \cdots + \gamma_n(n).$$

Now if $\alpha > 0$, then, for fixed n, the terms $\gamma_k(n)$ $(k = 1, 2, \ldots, n)$ of this sum increase until $\gamma_k(n) \leqq \gamma_{k+1}(n)$ or

$$k \leqq \xi = -\frac{\alpha+1}{2} + \sqrt{\left(\frac{\alpha+1}{2}\right)^2 + \alpha n}$$

If we denote the greatest term by μ_n, then it is attained in the sum for an index $k = p$ for which the estimate

$$p = \sqrt{\alpha n} + O(1)$$

is valid [1]. Since $\mu_n \leqq c_n \leqq n\mu_n$, we get

$$\log c_n = \log \mu_n + O(\log n).$$

Further,

$$\log \mu_n = \log \frac{n!}{p!(n-p)!} \cdot \frac{p}{n} \cdot \frac{\alpha^p}{p!}.$$

According to Stirling's formula, $\log m! = m \log m - m + O(\log m)$. We therefore obtain, finally,

$$\begin{aligned} \log c_n &= n \log n - n - p \log p + p - (n-p) \log (n-p) \\ &\quad + n - p + p \log \alpha - p \log p + p + O(\log n) \\ &= n \log n - p \log \frac{p^2}{\alpha} + p - (n-p) \left[\log n + \log \left(1 - \frac{p}{n}\right)\right] \\ &\quad + O(\log n) \\ &= 2p + O(\log n) = 2\sqrt{\alpha n} + O(\log n), \end{aligned}$$

which proves the assertion. In the last intermediate calculations we have to observe merely that $\frac{\log(1-z)+z}{z^2}$ approaches a limit as $z \to 0$.

---

[1] If $\{p_n\}$ is a sequence of positive terms, then $O(p_n)$ denotes any sequence whose terms, when divided by p_n, have *bounded* absolute values. In particular, $O(1)$ means a sequence which itself is bounded.

CHAPTER II

SINGULARITIES

§ 6. *The Laurent Expansion*

1. Let $\mathfrak{G}'$ be any closed region which, together with its boundary, lies in the annular region $\mathfrak{G}$. Then one can choose the closed paths $\mathfrak{C}_1$ and $\mathfrak{C}_2$ in $\mathfrak{G}$ in such a manner that $\mathfrak{C}_1$ encloses $\mathfrak{G}'$ as well as $\mathfrak{K}_2$, whereas $\mathfrak{C}_2$ encloses only $\mathfrak{K}_2$ and leaves $\mathfrak{G}'$ in its exterior — the proof follows K I, § 29 very closely. Then, for z in $\mathfrak{G}'$,

$$f(z) = \frac{1}{2\pi i}\int_{\mathfrak{C}_1} \frac{f(\zeta)}{\zeta - z}\,d\zeta - \frac{1}{2\pi i}\int_{\mathfrak{C}_2} \frac{f(\zeta)}{\zeta - z}\,d\zeta.$$

According to K I, § 16, each of the integrals represents a function which is regular in a neighborhood of every point which does not lie on $\mathfrak{C}_1$ or $\mathfrak{C}_2$, respectively. In particular, the first integral represents a regular function $f_1(z)$ inside $\mathfrak{C}_1$, and hence (!) actually inside $\mathfrak{K}_1$; the second, likewise, represents a regular function $f_2(z)$ outside $\mathfrak{K}_2$ (including ∞). Since both are regular in $\mathfrak{G}$,

$$f(z) = f_1(z) + f_2(z)$$

there. If we have similarly

$$f(z) = \varphi_1(z) + \varphi_2(z),$$

then

$$f_1(z) - \varphi_1(z) = \varphi_2(z) - f_2(z).$$

This function is regular inside $\mathfrak{K}_1$ and outside $\mathfrak{K}_2$ (including ∞), and hence in the entire plane (including ∞), so that it is a constant.

2. An argument entirely analogous to that employed in the preceding problem, where the region of regularity is now to be cut as in K I, p. 58, Fig. 4, leads to the representation

$$f(z) = f_1(z) + f_2(z) + \cdots + f_m(z),$$

where $f_1(z)$ denotes a function which is regular inside $\mathfrak{K}_1$, and each of the remaining $f_\nu(z)$ denotes a function which is regular outside $\mathfrak{K}_\nu$ and also at ∞ ($\nu = 2, \ldots, m$). If we have similarly

$$f(z) = \varphi_1(z) + \varphi_2(z) + \cdots + \varphi_m(z),$$

then from

$$f_1(z) - \varphi_1(z) = [\varphi_2(z) - f_2(z)] + \cdots + [\varphi_m(z) - f_m(z)]$$

it follows again that both sides of this equation are a constant c. From

$$f_2(z) - \varphi_2(z) = [\varphi_3(z) - f_3(z)] + \cdots + [\varphi_m(z) - f_m(z)] - c$$

it then follows analogously that $f_2(z) - \varphi_2(z)$ is a constant; etc.

3. a) Set $\frac{1}{z} = z'$. Then, for $|z| > 1$, $|z'| < 1$,

$$e^{\frac{1}{z-1}} = e^{\frac{z'}{1-z'}} = \sum_{n=0}^{\infty} c_n z'^n = \sum_{n=0}^{\infty} \frac{c_n}{z^n},$$

where the coefficients c_n have the same meaning as in § 5, Prob. 7 (with $\alpha = 1$).

b) For $|z| > 2$, $\sqrt{(z-1)(z-2)} = \pm z \left(1 - \frac{1}{z}\right)^{\frac{1}{2}} \left(1 - \frac{2}{z}\right)^{\frac{1}{2}}$.

Taking the binomial expansions here, our function for $|z| > 2$ is equal to

$$\pm z \left(1 - \binom{\alpha}{1}\frac{1}{z} + \binom{\alpha}{2}\frac{1}{z^2} - + \cdots\right)\left(1 - \binom{\alpha}{1}\frac{2}{z} + \binom{\alpha}{2}\frac{2^2}{z^2} - + \cdots\right)^{1)}$$

$$= \pm \left[c_0 z - c_1 + \frac{c_2}{z} - \frac{c_3}{z^2} + - \cdots\right],$$

if we set

$$c_n = \binom{\alpha}{n} + 2\binom{\alpha}{n-1}\binom{\alpha}{1} + 2^2\binom{\alpha}{n-2}\binom{\alpha}{2} + \cdots + 2^n\binom{\alpha}{n}.$$

c)

$$\frac{1}{(z-a)(z-b)} = \frac{1}{a-b}\left[\frac{1}{z-a} + \frac{1}{b-z}\right]$$

$$= \frac{1}{a-b}\left[\frac{1}{z}\frac{1}{1-\frac{a}{z}} + \frac{1}{b}\frac{1}{1-\frac{z}{b}}\right]$$

---

[1]) To facilitate printing we have set $\frac{1}{2} = \alpha$.

$$= \frac{1}{a - b}\left[\cdots + \frac{a^2}{z^3} + \frac{a}{z^2} + \frac{1}{z} + \frac{1}{b} + \frac{z}{b^2} + \frac{z^2}{b^3} + \cdots\right].$$

d) $$\frac{1}{(z-a)(z-b)} = \frac{1}{b-a}\left[\frac{b-a}{z^2} + \frac{b^2-a^2}{z^3} + \frac{b^3-a^3}{z^4} + \cdots\right].$$

e) $\log \frac{1}{1-z}$ does not have a Laurent expansion for $|z| > 1$, since the function is not single-valued there. If we set $\frac{1}{z} = z'$, then

$$\log \frac{-z'}{1-z'} = \log(-z') + \log \frac{1}{1-z'}$$
$$= \log\left(-\frac{1}{z}\right) + \frac{1}{z} + \frac{1}{2z^2} + \frac{1}{3z^3} + \cdots,$$

a substitute for the missing expansion.

4. The function $g(z)\cdot\gamma(z)$ is regular in $0 < |z| < \infty$ and is represented there by the series

$$\sum_{n=-\infty}^{+\infty} c_n z^n,$$

where

$$c_n = \sum_{k=-\infty}^{+\infty} a_k \alpha_{k-n}$$

for all $n \gtreqless 0$. Here all a_ν and α_ν with a negative index are to be set equal to 0. (What guarantees the convergence of these series? Cf. the next solution.)

5. Let $\sum_{n=0}^{\infty} \alpha_n$ and $\sum_{n=0}^{\infty} \beta_n$ be any two absolutely convergent series (cf. § 2, Prob. 2a). Then every series $\sum_{n=0}^{\infty} \gamma_n$, in which the γ_n denote any of the products $\alpha_p\beta_q$, but not the same one of these products for distinct n, converges. If $\sum \gamma_n$ contains *all* products $\alpha_p\beta_q$, its value is $\sum\alpha_n \sum\beta_n$. Now, for a z in the interior of the ring of convergence, the given series are absolutely convergent, and hence, according to the remark just made, not only the series

$$c_n = \sum_{k=-\infty}^{+\infty} a_k b_{n-k} = z^{-n} \sum_{k=-\infty}^{+\infty} (a_k z^k)(b_{n-k} z^{n-k}),$$

$$n = 0, \pm 1, \pm 2, \ldots,$$

but also the series

$$\sum_{n=-\infty}^{+\infty} c_n z^n$$

formed with these coefficients, which series thus represents the Laurent expansion of the product.

(It makes no essential difference that the series are infinite in *both* directions, since one can also think of every term with a negative index as being placed after the corresponding term having a positive index.)

6. $$e^{c\left(z+\frac{1}{z}\right)} = e^{cz} \cdot e^{\frac{c}{z}}$$

$$= \left(1 + cz + \frac{c^2}{2!} z^2 + \cdots\right) \left(1 + \frac{c}{z} + \frac{c^2}{2!} \frac{1}{z^2} + \cdots\right)$$

$$= \sum_{n=-\infty}^{+\infty} c_n z^n = c_0 + \sum_{n=1}^{\infty} c_n \left(z^n + \frac{1}{z^n}\right),$$

if we set $c_n = c_{-n} = \sum_{k=0}^{\infty} \frac{c^{n+k}}{(n+k)!} \cdot \frac{c^k}{k!}$, $n \geqq 0$.

§ 7. *The Various Types of Singularities*

1. a) $= (z^2 + 4) e^{-z}$; essential singularity.

b) $= \pm z \left(1 - \frac{1}{z}\right)^{\frac{1}{2}} \left(1 - \frac{2}{z}\right)^{\frac{1}{2}} = \pm z \left(1 + \frac{a_1}{z} + \frac{a_2}{z^2} + \cdots\right)$; on each of the two sheets, which are unattached at ∞ and for a neighborhood thereof, there is a simple pole.

c) $1 - z - \frac{z^2}{2!} + \frac{z^3}{3!} + - \cdots$; essential singularity.

d) and e) do not have an isolated singularity at ∞, and therefore do not come under the classification in K I, § 31. The point ∞ here is a limit point of simple poles.

f) $= 1 - \frac{1}{z^2} + \frac{1}{2!z^4} - + \cdots$; regular and equal to $+1$ at ∞.

g) Set $1/z = z'$; then, for $|z| > 1$, $|z'| < 1$,

$$\sin \frac{1}{1-z} = -\sin \frac{z'}{1-z'} = -\frac{z'}{1-z'} + \frac{1}{3!}\left(\frac{z'}{1-z'}\right)^3 - + \cdots,$$

and hence (cf. I, § 10, 1b),

$$= -z' + a_2 z'^2 + \cdots = -\frac{1}{z} + \frac{a_2}{z^2} + \cdots.$$

The function is thus regular at ∞ and has a simple zero there.

h) and i) Here ∞ is again a limit point of simple poles.

2. a) $= 1 + \frac{1}{z} + \frac{1}{2!z^2} + \cdots$; essential singularity.

b) $= -\frac{1}{(z-1)} + \frac{1}{3!(z-1)^3} - + \cdots$; essential singularity.

c) $$= \frac{1}{1 - \left[1 + (z - 2\pi i) + \frac{1}{2!}(z - 2\pi i)^2 + \cdots\right]}$$

$= -\frac{1}{z - 2\pi i} + \frac{1}{2} + a_1(z - 2\pi i) + \cdots$; simple pole with residue -1.

d) $\sin z - \cos z = \sqrt{2} \sin\left(z - \frac{\pi}{4}\right) = \sqrt{2}\left(z - \frac{\pi}{4}\right) - \frac{\sqrt{2}}{6}\left(z - \frac{\pi}{4}\right)^2$ $+ \cdots$; simple pole with residue $\frac{1}{2}\sqrt{2}$.

3. $f_1 \pm f_2$ also has a pole of order β; $f_1 f_2$ has a pole of order $(\beta - \alpha)$, or a zero of order $(\alpha - \beta)$, or a regular point with a nonzero functional value, according as $\beta > \alpha$, $\beta < \alpha$, or $\beta = \alpha$; $\frac{f_1}{f_2}$ has a zero of order $(\alpha + \beta)$; $\frac{f_2}{f_1}$ has a pole of order $(\alpha + \beta)$. The proofs are effected by forming the power series.

4. Since e^z is an entire function, there exist arbitrarily large circles

$\mathfrak{K}$ about 0 for which the assumptions of Theorem 2 in K I, § 33 are fulfilled. The number of zeros lying inside $\mathfrak{K}$ is equal to

$$\frac{1}{2\pi i}\int_{\mathfrak{K}} dz = 0.$$

5. The real function possesses a zero at $x = 0$, is differentiable there, and actually has derivatives there of every order, all of which have the value 0. The graph does not indicate any irregular behavior at $x = 0$, and is at most remarkable for the fact that the order of contact of the curve and the x-axis is greater than that of *any* parabola $y = x^n$. The function of a complex variable, $w = e^{-\frac{1}{z^2}}$, however, has an essential singularity at $z = 0$, so that it is neither continuous nor differentiable there, and in every neighborhood of 0 the function comes arbitrarily close to every value.

6. $F_0(z)$ has no region of definition at all; for, the integral, which is improper at z_0, diverges because $\beta \geqq 1$.

To form $F_1(z)$, the Laurent expansion of $f(z)$ may be integrated term by term. We see then immediately that $F_1(z)$ is single-valued in a neighborhood of z_0 if, and only if, the residue of $f(z)$ at z_0 is equal to 0; in particular, therefore, $\beta \geqq 2$. $F_1(z)$ then has a pole of order $(\beta - 1)$ at z_0. If the residue mentioned is equal to $c \neq 0$, then $F_1(z)$ has a logarithmic singularity at z_0; more precisely: $F_1(z) - c \log (z - z_0)$ is single-valued in a neighborhood of z_0 and has a pole of order $(\beta - 1)$ at z_0 (or is regular there if $\beta = 1$).

7. The expansion

$$\frac{\varphi(\zeta)}{\zeta - z} = -\frac{1}{z}\varphi(\zeta) - \frac{1}{z^2}\cdot\zeta\varphi(\zeta) - \frac{1}{z^3}\cdot\zeta^2\varphi(\zeta) - \cdots,$$

for fixed, sufficiently large z, converges uniformly with respect to ζ along $\mathfrak{k}$, so that the Laurent expansion of $f(z)$, valid outside $\mathfrak{K}$, has the form $\frac{a_1}{z} + \frac{a_2}{z^2} + \cdots$. Thus, $f(z)$ is regular at ∞ and has a zero there.

8. Let $f(z) = \sum_{n=-\infty}^{+\infty} a_n z^n$ for $|z| > R$. This expansion can be integrated

term by term along $\mathfrak{k}$, each term yielding a unique value (i.e., one that is independent of $\mathfrak{k}$) except

$$a_{-1} \cdot \int_{z_0,\,\mathfrak{k}}^{z} \frac{d\zeta}{\zeta}.$$

Hence, $F(z)$ is single-valued if, and only if, $a_{-1} = 0$. If this condition is satisfied, and if $f(z)$ has a pole of order β at ∞, then $F(z)$ has a pole of order $(\beta + 1)$, and, in particular, a simple pole if $f(z)$ is regular and $\neq 0$ at ∞. If $f(z)$ has a zero of order α at ∞ ($\alpha \geqq 2$ because $a_{-1} = 0$), then $F(z)$ has a zero of order $(\alpha - 1)$. If $a_{-1} \neq 0$, then $F(z) - a_{-1} \cdot \log z$ remains regular for $|z| > R$ and again exhibits an easily described behavior at ∞.

9. Let $z_0 \neq \infty$ and let $f(z)$ be bounded in a neighborhood of z_0. Since the second one of the integrals written down in the problem has the same value for all sufficiently small circles $\mathfrak{K}_2$ (cf. K I, p. 119, center), to find this value one may let the radius ρ of $\mathfrak{K}_2$ decrease to zero. Now if $\rho < |z - z_0|$ and M denotes an upper bound for $|f(z)|$, then

$$\left| \frac{1}{2\pi i} \int_{\mathfrak{K}_2} \frac{f(\zeta)}{\zeta - z}\, d\zeta \right| \leqq \frac{1}{2\pi} \cdot 2\pi\rho \cdot \frac{M}{|z - z_0| - \rho},$$

which decreases to 0 with ρ. Therefore this second integral equals 0, and $f(z)$ is equal to the first integral. The latter, however, according to K I, § 16, defines a function which is regular in $\mathfrak{K}_1$ and hence, in particular, at z_0. Since this function coincides with $f(z)$ for $z \neq z_0$, $f(z)$ is regular at z_0. Conversely, since a function which is regular at z_0 is bounded in a neighborhood of z_0, the first part of Riemann's theorem is proved for $z_0 \neq \infty$. The case $z_0 = \infty$ is reduced to the case $z_0 = 0$ by means of the transformation $z = \frac{1}{z'}$. The other two parts of Riemann's theorem now result once more immediately from the definition of a pole or of an essential singular point as a point on approaching which the function becomes definitely infinite or completely indefinite, respectively.

10. Suppose that there were a point z_0, a number c, and an $\varepsilon > 0$, such that $f(z)$ was single-valued and regular in a neighborhood of z_0 (excluding z_0 itself), but invariably $|f(z) - c| \geqq \varepsilon$. Then $\frac{1}{f(z) - c}$ would also be single-valued and regular, as well as bounded, there. According to Riemann's theorem, z_0 would have to be a regular point of this function. Consequently, $f(z)$ would have a regular point or a pole there, contrary to assumption.

11. Since the proofs for $z_0 = \infty$ and $z_0 \neq \infty$ are quite analogous, it is sufficient to consider the first case. Let c be an arbitrary complex number, and let ε be an arbitrarily small, R an arbitrarily large, positive number. Call the circle with radius ε about c, $\mathfrak{C}$, and the circle with radius R about 0, $\mathfrak{K}$. Then, according to Casorati-Weierstrass, there exists a z_1 outside $\mathfrak{K}$ such that $f(z_1) = c_1$ lies inside $\mathfrak{C}$. We can describe a circle $\mathfrak{K}_1$ about z_1 and a circle $\mathfrak{C}_1$ about c_1 such that all values in $\mathfrak{C}_1$ are assumed by $f(z)$ in $\mathfrak{K}_1$ (cf. B, p. 6, as well as K I, § 34). Here we suppose $\mathfrak{C}_1$ to be so small that it lies entirely within $\mathfrak{C}$, and $\mathfrak{K}_1$ so small that it lies entirely without $\mathfrak{K}$. Now there exists a $|z_2| > 2R$ such that $f(z_2) = c_2$ lies inside $\mathfrak{C}_1$. We describe $\mathfrak{K}_2$ about z_2 and the circle C_2 about c_2 such that all values in $\mathfrak{C}_2$ are assumed by $f(z)$ in $\mathfrak{K}_2$. Moreover, let $\mathfrak{C}_2$ lie wholly within $\mathfrak{C}_1$, and $\mathfrak{K}_2$ lie wholly without $\mathfrak{K}$ and $\mathfrak{K}_1$. Now there exists a $|z_3| > 3R$ such that $f(z_3) = c_3$ lies inside $\mathfrak{C}_2$; etc. There is at least one point a which is common to all the circles $\mathfrak{C}, \mathfrak{C}_1, \mathfrak{C}_2, \mathfrak{C}_3, \ldots$. The equation $f(z) = a$, then, has at least one solution in each of the circles $\mathfrak{K}_1, \mathfrak{K}_2, \ldots$, and hence infinitely many solutions outside $\mathfrak{K}$, Q.E.D. (Draw the circles $\mathfrak{K}, \mathfrak{K}_1, \ldots$ in the z-plane, the circles $\mathfrak{C}, \mathfrak{C}_1, \ldots$ in a w-plane.)

§ 8. *The Residue Theorem, Zeros, and Poles*

1. We have

$$\frac{\varphi(z)}{f(z)} = \frac{b_0}{z^n} + \frac{b_1}{z^{n-1}} + \cdots + \frac{b_{n-1}}{z},$$

if we set $\frac{a_\nu}{a_n} = b_\nu \quad (\nu = 0, 1, \cdots, n-1)$.

For $|z| > 1$, therefore, $\left|\dfrac{\varphi(z)}{f(z)}\right| < \dfrac{|b_0| + |b_1| + \cdots + |b_{n-1}|}{|z|}$,

and hence < 1 for $|z| = R$, provided that $R > 1$ is chosen large enough. This R, moreover, according to K I, § 28, Theorem 2, can be chosen so large that the polynomial $a_0 + a_1 z + \cdots + a_n z^n$ certainly has no zero outside the circle $|z| = R$ or on its boundary. The theorem mentioned in the problem is now applicable to the circle $|z| = R$. Since $f(z) = a_n z^n$ has only the zero 0 of order n in $|z| < R$, $f(z) + \varphi(z) = a_0 + a_1 z + \cdots + a_n z^n$ also has precisely n zeros in $|z| < R$, and no others.

2. ze^{z^2} or $(z^2 - 1)e^{cz}$, if c is not real.

3. If $f(z)$ has a *simple* pole at z_0, then the residue there $= \lim\limits_{z \to z_0} (z - z_0) f(z)$. Otherwise one has to set up the Laurent expansion to obtain the residue.

a) $= \lim\limits_{z \to k\pi} \dfrac{(z - k\pi)}{\sin z} = \left(\dfrac{1}{\cos z}\right)_{z = k\pi} = (-1)^k.$

b₁) $= \lim\limits_{z \to 1} \dfrac{z}{(z-2)^2} = 1.$

b₂) $f(z) = \dfrac{1}{(z-2)^2} \cdot \dfrac{2 + (z-2)}{1 + (z-2)}$

$$= \frac{2}{(z-2)^2} - \frac{1}{z-2} + c_0 + c_1(z-2) + \cdots .$$

The residue $= -1$.

c₁) $f(z) = \dfrac{1}{(z-z_1)^m} \cdot \dfrac{-a}{z_2 - z_1} \left[1 + \dfrac{z - z_1}{z_2 - z_1} + \left(\dfrac{z - z_1}{z_2 - z_1}\right)^2 + \cdots\right].$

The residue $= \dfrac{-a}{(z_2 - z_1)^m}$.

c₂) $= \lim\limits_{z \to z_2} \dfrac{a}{(z - z_1)^m} = \dfrac{a}{(z_2 - z_1)^m}$.

d) $= \lim_{z \to z_k} \dfrac{(z - z_k) \sin z}{\cos z} = -1.$

e) $e^{\frac{1}{z}} = 1 + \dfrac{1}{z} + \dfrac{1}{2!} \cdot \dfrac{1}{z^2} + \cdots.$ The residue $= 1$.

f) $e^{\frac{1}{z-1}} = 1 + \dfrac{1}{z-1} + \cdots.$ The residue $= 1$.

g) $= \lim_{z \to 2k\pi i} \dfrac{z - 2k\pi i}{1 - e^z} = \left(\dfrac{1}{-e^z}\right)_{z = 2k\pi i} = -1.$

4. It can be read off from

$$[\varphi(z_0) + \varphi'(z_0) \cdot (z - z_0) + \cdots] \left[\frac{\alpha}{z - z_0} + \cdots\right].$$

The residue $= \alpha z_0$, $\alpha\varphi(z_0)$, respectively. Analogously for the second question: $-\beta z_0$, $-\beta \cdot \varphi(z_0)$, respectively.

5. According to K I, § 33, Theorem 3 and by the preceding problem, under the assumptions of the theorem mentioned, the first integral

$$= \Sigma z_\nu - \Sigma z'_\mu$$

extended over the zeros z_ν lying within $\mathfrak{C}$ or the poles z'_μ there, each taken as often as the corresponding order indicates. Analogously the second integral

$$= \Sigma \varphi(z_\nu) - \Sigma\varphi(z'_\mu).$$

6. If we set

$$\frac{1}{\nu!} f^{(\nu)}(z_0) = a_\nu, \quad \frac{1}{\nu!} g^{(\nu)}(z_0) = b_\nu, \quad \nu = 0, 1, 2, \cdots,$$

then

$$\frac{g(z)}{f(z)} = \frac{a_0 + a_1(z - z_0) + a_2(z - z_0)^2 + \cdots}{b_2(z - z_0)^2 + b_3(z - z_0)^3 + \cdots}; \quad a_0 \neq 0, \quad b_2 \neq 0;$$

$$= \frac{1}{b_2 z'^2} \left[a_0 + a_1 z' + \cdots\right] \left[1 - \frac{b_3}{b_2} z' + \cdots\right]; \quad z' = (z - z_0).$$

The residue $= \dfrac{a_1 b_2 - a_0 b_3}{b_2^2}.$

For the second question one has analogously for $b_3 \neq 0$:

$$\frac{f(z)}{g(z)} = \frac{1}{b_3 z'^3}\left[a_0 + a_1 z' + a_2 z'^2 + \cdots\right]\left[1 - \frac{b_4}{b_3} z' + \left(\frac{b_4^2}{b_3^2} - \frac{b_5}{b_3}\right) z'^2 + \cdots\right].$$

$$\text{The residue} = \frac{a_0 b_4^2 - a_0 b_3 b_5 - a_1 b_3 b_4 + a_2 b_3^2}{b_3^3}.$$

7. a) $= 2\pi i$.

b) Since all poles (which lie at the odd multiples of $\frac{1}{2}$) have the residue $-\frac{1}{\pi}$ (cf. Prob. 3d), the integral $= -4ni$.

c) $$= 2\pi i \left(\frac{f(z_1)}{(z_1 - z_2)\cdots(z_1 - z_k)} + \frac{f(z_2)}{(z_2 - z_1)\cdots(z_2 - z_k)} + \cdots + \frac{f(z_k)}{(z_k - z_1)\cdots(z_k - z_{k-1})}\right).$$

8. From the assumptions it follows immediately that r_0 can be chosen so large that the numerator and denominator of $R(z)$ have no zero for $|z| > r_0$, and that, for a suitable $\alpha > 0$, $|R(z)| < \frac{\alpha}{|z|^2}$ there. This implies that the integral converges. Let $r > r_0$, and let $\mathfrak{C}$ be the closed path leading from $-r$ along the axis of reals to $+r$ and thence along the upper half $\mathfrak{H}$ of the circle $|z| = r$ back to $-r$. Then

$$\int_{\mathfrak{C}} R(z)\,dz = \int_{-r}^{+r} R(x)\,dx + \int_{\mathfrak{H}} R(z)\,dz = 2\pi i S.$$

Here, however,

$$\left|\int_{\mathfrak{H}} R(z)\,dz\right| \leq \frac{\alpha}{r^2} \cdot \pi r = \frac{\alpha\pi}{r}.$$

This yields the assertion (and once more the convergence of the integral) as $r \to +\infty$.

9. According to Cauchy's integral theorem,

$$I_1 + I_2 + I_3 + I_4 = 0,$$

and hence

$$I_1 + I_3, \text{ or what is the same, } 2i \int_{\rho}^{r} \frac{\sin x}{x}\, dx = -I_2 - I_4.$$

According to I, § 7, Prob. 6a and b, $I_2 \to 0$ as $r \to \infty$ and $I_4 \to -\pi i$ as $\rho \to 0$. Hence,

$$\int_0^{\infty} \frac{\sin x}{x}\, dx = \frac{\pi}{2}.$$

10. We have $\int_{\mathfrak{C}} e^{-z^2} dz = 0$, and consequently

$$0 = \int_0^R e^{-t^2}\, dt + \int_0^{\frac{1}{4}\pi} e^{-R^2 (\cos 2\varphi + i \sin 2\varphi)} \cdot iRe^{i\varphi} d\varphi + \int_{z_1}^{0} e^{-z^2} dz,$$

where the last integral is again taken rectilinearly. As $R \to +\infty$, the first integral has the limit $\frac{1}{2}\sqrt{\pi}$; and the second has the limit 0, because in absolute value it is

$$\leqq R \int_0^{\frac{\pi}{4}} e^{-R^2 \cos 2\varphi} d\varphi = \tfrac{1}{2} R \int_0^{\frac{\pi}{2}} e^{-R^2 \sin \psi} d\psi, \qquad \left(\varphi = \frac{\pi}{4} - \frac{\psi}{2}\right).$$

Now, as is well known, $\sin \varphi > \frac{2}{\pi}\varphi$ in $0 \ldots \frac{\pi}{2}$, so that the integral is also

$$< \tfrac{1}{2} R \cdot \int_0^{\frac{\pi}{2}} e^{-\frac{2}{\pi} R^2 \varphi} d\varphi < \frac{1}{R},$$

and thus tends to 0. For $z = \dfrac{1+i}{\sqrt{2}}\, t$, our third integral

$$= -\frac{1+i}{\sqrt{2}} \int_0^R e^{-it^2} dt = -\frac{1+i}{\sqrt{2}} \left(\int_0^R \cos(t^2) dt - i \int_0^R \sin(t^2) dt \right).$$

Therefore, as $R \to \infty$,

$$\int_0^R \cos(t^2)\,dt - i\int_0^R \sin(t^2)\,d \to \frac{\sqrt{2}}{1+i}\cdot\tfrac{1}{2}\sqrt{\pi} = \tfrac{1}{2}\sqrt{\frac{\pi}{2}}\cdot(1-i).$$

Separation into real and imaginary parts yields the two formulas.

ENTIRE AND MEROMORPHIC FUNCTIONS

§ 9. *Infinite Products. Weierstrass's Factor-theorem*

1. According to K II, § 2, Theorems 4 and 5, the products are absolutely convergent. We have, further,

$$\text{a)} \quad = \prod_{n=1}^{\infty} \frac{(n+1)^2}{n(n+2)}$$

$$= \lim_{n\to\infty} \frac{2\cdot 2}{1\cdot 3} \cdot \frac{3\cdot 3}{2\cdot 4} \cdot \ \cdots \ \cdot \frac{(n+1)\,(n+1)}{n(n+2)} = 2.$$

$$\text{b)} \quad = \prod_{n=2}^{\infty} \frac{(n-1)\,(n+2)}{n\,(n+1)}$$

$$= \lim_{n\to\infty} \frac{1\cdot 4}{2\cdot 3} \cdot \frac{2\cdot 5}{3\cdot 4} \cdot \ \cdots \ \cdot \frac{(n-1)\,(n+2)}{n(n+1)} = \tfrac{1}{3}.$$

$$\text{c)} \quad = \prod_{n=2}^{\infty} \frac{(n-1)\,[(n+1)^2-(n+1)+1]}{(n+1)\,(n^2-n+1)}$$

$$= \lim_{n\to\infty} \frac{1\cdot 2}{3} \cdot \frac{n^2+n+1}{n(n-1)} = \tfrac{2}{3}.$$

$$\text{d)} \quad \prod_{n=1}^{\infty} \left(1+\frac{1}{n^2}\right) = \frac{\sin \pi z}{\pi z} \text{ for } z = i, \text{ and hence } = \frac{e^{\pi}-e^{-\pi}}{2\pi}.$$

(Cf. § 5, Prob. 5.)

2. Subtract the nth partial sum of the series from 1. By collecting terms step by step, we obtain

$$(1-\theta_1)\ (1-\theta_2) \cdots (1-\theta_n).$$

Since this is >0, the partial sums are <1, so that the series converges. More precisely: If $\Sigma\theta_n$ is divergent, then the given series converges to the sum 1. If $\sum\theta_n$ is convergent, then, according to K II, § 2, Theorem 5, so is $\prod(1-\theta_n)$, and the given series converges to the sum $1-\prod(1-\theta_n)$.

3. a) $e^{\frac{1}{2}x} \leqq 1+x \leqq e^x$ in $0 \leqq x \leqq 1$; cf. I, § 6, Prob. 3. Under each of the assumptions, γ_ν must $\to 0$. Hence, for all λ after a certain one, and all $k \geqq 1$,

$$e^{\frac{1}{2}(\gamma_\lambda+\gamma_{\lambda+1}+\cdots+\gamma_{\lambda+k})} \leqq \prod_{\nu=\lambda}^{\lambda+k}(1+\gamma_\nu) \leqq e^{(\gamma_\lambda+\gamma_{\lambda+1}+\cdots+\gamma_{\lambda+k})}.$$

According to the right half of this inequality, the boundedness of the partial sums of $\sum\gamma_\nu$ implies that of the partial products of $\prod(1+\gamma_\nu)$, and the converse is a consequence of the left half.

b) We have

$$\gamma_1+\gamma_2+\cdots+\gamma_n < (1+\gamma_1)(1+\gamma_2)\cdots(1+\gamma_n).$$

The boundedness of the partial products thus implies that of the partial sums. On the other hand, if $\sum\gamma_\nu$ converges, and if we choose m so that $\gamma_{m+1}+\gamma_{m+2}+\cdots+\gamma_{m+k} < \frac{1}{2}$ for all $k \geqq 1$, then it follows that

$$(1+\gamma_{m+1})\cdots(1+\gamma_{m+k}) < 1+(\gamma_{m+1}+\cdots+\gamma_{m+k}) + (\gamma_{m+1}+\cdots+\gamma_{m+k})^2+\cdots,$$

and hence < 2 for all $k \geqq 1$. Therefore $\prod(1+\gamma_\nu)$ is also convergent.

4. From the assumption it follows first that $a_n \to 0$. Now, for $|z| < \frac{1}{2}$, obviously $|\operatorname{Log}(1+z)-z| \leqq \frac{|z|^2}{2}(1+|z|+|z|^2+\cdots) < |z|^2$. Choose m so that invariably $|a_n| < \frac{1}{2}$ for $n > m$. Then, for all $n > m$ and all $k \geqq 1$,

$$\left|\sum_{\nu=n+1}^{n+k}\operatorname{Log}(1+a_\nu) - \sum_{\nu=n+1}^{n+k}a_\nu\right| \leqq \sum_{\nu=n+1}^{n+k}|a_\nu|^2,$$

which $\to 0$ as $n \to \infty$. From this one can read off the assertion. The assertion concerning the absolute convergence of the series follows immediately from $\left|\frac{\operatorname{Log}(1+a_n)}{a_n}\right| \to 1$.

5. a) and b) $|z|<1$;

c) the entire plane;

d) and e) the half-plane $\Re(z)>1$. Proof, in each case, according to K II, § 2, Theorem 5.

6. a) $(1-z)\cdot(1+z)(1+z^2)\cdots(1+z^{2^n})=1-z^{2^{n+1}}$, which $\to 1$.

b) According to a), $\dfrac{1}{1-z^{2\nu+1}}=\prod\limits_{n=0}^{\infty}(1+z^{2^n(2\nu+1)})$. The exponents $2^n(2\nu+1)$, however, yield, for $n=0,1,2,\ldots$, $\nu=0,1,2,\ldots$, every natural number once and only once.

7. Since a finite number of power series may be multiplied according to the elementary rules, it is at any rate permissible to write (*) in the statement of the problem; i.e., $A_\lambda^{(n)}$ (for fixed n and λ) is obtained by suppressing all higher powers $z^{\lambda+1}$, $z^{\lambda+2}$, ... on the left, multiplying out the finite product of ordinary sums $\prod\limits_{\nu=1}^{n}(1+a_0^{(\nu)}+a_1^{(\nu)}z+\cdots+a_\lambda^{(\nu)}z^\lambda)$, and collecting the terms involving z^λ. According to K II, p. 13, the series $P_1+(P_2-P_1)+\cdots$ converges uniformly in $|z|\leqq\rho<r$. Hence, by Weierstrass's double-series theorem (K I, § 20, Theorem 3), for fixed λ, the series

$$A_\lambda^{(1)}+(A_\lambda^{(2)}-A_\lambda^{(1)})+\cdots=\lim_{n\to\infty}A_\lambda^{(n)}$$

is also convergent and its value is the coefficient A_λ.

8. We have $\dfrac{\sin\pi z}{\pi z}=1-\dfrac{\pi^2z^2}{6}+\dfrac{\pi^4z^4}{120}-+\cdots$, and, on the other hand, by multiplying out the product according to the preceding problem,

$$=1-\left(\sum_{n=1}^{\infty}\frac{1}{n^2}\right)z^2+\left[\sum_{k=1}^{\infty}\left(\sum_{n=k+1}^{\infty}\frac{1}{k^2n^2}\right)\right]z^4-+\cdots.$$

Hence, $\sum\limits_{n=1}^{\infty}\dfrac{1}{n^2}=\dfrac{\pi^2}{6}$ and

$$\sum_{k=1}^{\infty}\frac{1}{k^2}\left(\sum_{n=k+1}^{\infty}\frac{1}{n^2}\right)=\sum_{n=2}^{\infty}\frac{1}{n^2}\left(\sum_{k=1}^{n-1}\frac{1}{k^2}\right)=\frac{\pi^4}{120}.$$

If we add the last two series term by term, we obtain

$$\left(1 + \frac{1}{2^2} + \frac{1}{3^2} + \cdots\right)^2 - \left(1 + \frac{1}{2^4} + \frac{1}{3^4} + \cdots\right) = \frac{\pi^4}{60}.$$

Consequently, $\sum_{n=1}^{\infty} \frac{1}{n^4} = \frac{\pi^4}{36} - \frac{\pi^4}{60} = \frac{\pi^4}{90}$.

9. a) $C_0 = 1, \quad C_1 = \sum_{\nu=1}^{\infty} c_\nu, \quad C_2 = \sum_{1 \leqq \varkappa < \lambda} c_\varkappa c_\lambda,$

$$C_3 = \sum_{1 \leqq \varkappa < \lambda < \mu} c_\varkappa c_\lambda c_\mu, \cdots.$$

b) $A_0 = 1, \quad A_1 = \sum_{\nu=1}^{\infty} q^{2\nu-1} = \frac{q}{1 - q^2},$

$$A_2 = \frac{q^4}{(1 - q^2)\,(1 - q^4)}, \cdots.$$

A_1 and A_2 are found directly according to a), the general form of A_ν is found more conveniently as follows: We have

$$\Pi\,(1 + q^{2n-1}z) = f(z) = (1 + qz)\,f(q^2z),$$

and hence

$$1 + A_1z + A_2z^2 + \cdots = (1 + qz)\,(1 + A_1q^2z + A_2q^4z^2 + \cdots).$$

This yields

$$A_\nu = A_{\nu-1} \cdot \frac{q^{2\nu-1}}{1 - q^{2\nu}}, \text{ which implies } A_\nu = \frac{q^{\nu^2}}{(1 - q^2)\,(1 - q^4) \cdots (1 - q^{2\nu})}.$$

c) According to a), $\sum_{\nu=-\infty}^{+\infty} D_\nu z^\nu = \sum_{\nu=0}^{\infty} C_\nu z^\nu \cdot \sum_{\nu=0}^{\infty} \frac{C_\nu}{z^\nu}$. Hence, $D_0 = C_0^2 + C_1^2 + C_2^2 + \cdots$, and in general, for all $\nu \geqq 0$, $D_\nu = D_{-\nu} = C_0C_\nu + C_1C_{\nu+1} + C_2C_{\nu+2} + \cdots$. The expansion therefore can also be written in the form $D_0 + D_1\left(z + \frac{1}{z}\right) + D_2\left(z^2 + \frac{1}{z^2}\right) + \cdots$ (Cf. § 6, Prob. 5).

d) According to b) and c), $B_\nu = B_{-\nu} = A_0A_\nu + A_1A_{\nu+1} + \cdots,$

$\nu \geqq 0$. Values in closed form are obtained as follows: Since $F(z) = qz \cdot F(q^2 z)$, we have

$$B_0 + B_1\left(z + \frac{1}{z}\right) + \cdots = B_0 qz + \left(B_1 q^3 z^2 + \frac{B_1}{q}\right) + \cdots.$$

Comparison yields $B_\nu = q^{2\nu-1} B_{\nu-1}$, $\nu = 1, 2, \ldots$, and consequently $B_\nu = q^{\nu^2} \cdot B_0$.

Therefore

$$F(z) = B_0 \cdot \left[1 + q\left(z + \frac{1}{z}\right) + \cdots + q^{\nu^2}\left(z^\nu + \frac{1}{z^\nu}\right) + \cdots\right].$$

B_0 is determined from $B_\nu = q^{\nu^2} \cdot B_0 = A_\nu + A_1 A_{\nu+1} + \cdots$. For according to this, if we set $(1 - q^2)(1 - q^4) \cdots (1 - q^{2\nu}) = p_\nu$ for brevity,

$$q^{\nu^2} \cdot B_0 = \frac{q^{\nu^2}}{p_\nu} + \frac{q \cdot q^{(\nu+1)^2}}{p_1 \cdot p_{\nu+1}} + \cdots,$$

and hence

$$|p_\nu B_0 - 1| \leqq \frac{|q|^{2\nu} + |q|^{4\nu} + |q|^{6\nu} + \cdots}{[(1 - |q|^2)(1 - |q|^4) \cdots]^2} \leqq c \cdot |q|^{2\nu},$$

where c denotes a suitable positive number. Consequently, $p_\nu B_0 \to 1$, i.e., $B_0 = \prod_{\nu=1}^{\infty} \frac{1}{1 - q^{2\nu}}$. Thus,

$$\prod_{n=1}^{\infty}(1 - q^{2n}) \cdot \prod_{n=1}^{\infty}(1 + q^{2n-1} z) \prod_{n=1}^{\infty}\left(1 + \frac{q^{2n-1}}{z}\right) = 1 + \sum_{\nu=1}^{\infty} q^{\nu^2}\left(z^\nu + \frac{1}{z^\nu}\right)$$

is the desired expansion.

e) In d), replace z by $-z^{1/2}$ and at the same time q by $z^{3/2}$. Then, for $|z| < 1$, the left-hand side

$$= \prod_{n=1}^{\infty}(1 - z^{3n}) \cdot \prod_{n=1}^{\infty}(1 - z^{3n-1}) \cdot \prod_{n=1}^{\infty}(1 - z^{3n-2}).$$

Therefore

$$\prod_{n=1}^{\infty}(1 - z^n) = 1 + \sum_{\nu=1}^{\infty}(-1)^\nu\left(z^{\frac{3\nu^2-\nu}{2}} + z^{\frac{3\nu^2+\nu}{2}}\right)$$

$$= 1 - z - z^2 + z^5 + z^7 - z^{12} - z^{15} + + - - \cdots.$$

(*Euler-Legendre theorem on pentagonal numbers.*)

10. a) If, in $\frac{\sin \pi z}{\pi z} = \Pi\left(1-\frac{z^2}{n^2}\right)$, we set first $z = \frac{1}{2}$ and then $z = \frac{1}{4}$, and divide the second result by the first, we obtain the assertion immediately because $\Pi\left(1-\frac{1}{(2n)^2}\right) = \Pi\left(1-\frac{1}{(4k-2)^2}\right)\left(1-\frac{1}{(4k)^2}\right)$.

b) Analogously for $z = \frac{1}{6}$ and $z = \frac{1}{3}$.

11. a)
$$e^z - 1 = 2i \cdot e^{\frac{z}{2}} \cdot \frac{e^{i\frac{z}{2i}} - e^{-i\frac{z}{2i}}}{2i} = 2ie^{\frac{z}{2}} \cdot \sin\frac{z}{2i}$$
$$= e^{\frac{z}{2}} \cdot z \cdot \prod_{n=1}^{\infty}\left(1+\frac{z^2}{4\pi^2 n^2}\right).$$

b) Because of a) we may assume that $z_0 \neq 2k\pi i$, $k = 0, \pm 1, \ldots$. By a),
$$e^z - e^{z_0} = e^{z_0}\left(e^{z-z_0} - 1\right)$$
$$= z_0 \cdot e^{z_0} \cdot e^{\frac{z-z_0}{2}}\left(1 - \frac{z}{z_0}\right)\prod_{n=1}^{\infty}\left(1+\frac{(z-z_0)^2}{4n^2\pi^2}\right)$$
is a product representation of the given function, which accomplishes what is required of the Weierstrass product. It does not, however, have the exact form of a Weierstrass product (see K II, § 2, p. 18). In order to obtain it, the product in the last formula must be recast as follows:
$$\Pi = \prod_{n=1}^{\infty}\left\{\left(1+\frac{z-z_0}{2n\pi i}\right)\left(1-\frac{z-z_0}{2n\pi i}\right)\right\}$$
$$= \prod_{n=1}^{\infty}\left\{\left(1+\frac{z_0}{2n\pi i}\right)\left(1-\frac{z_0}{2n\pi i}\right)\right\}\cdot\prod_{n=1}^{\infty}\left\{\left(1-\frac{z}{z_0+2n\pi i}\right)\left(1-\frac{z}{z_0-2n\pi i}\right)\right\}.$$

If one wishes to remove the braces, one must introduce the convergence-producing factors.

c)
$$\cos \pi z = \frac{\sin 2\pi z}{2 \sin \pi z} = \prod_{n=1}^{\infty}\left(1-\frac{4z^2}{(2n-1)^2}\right).$$

d) and e), because of

$$\sin \pi z - \sin \pi z_0 = 2 \cos \pi \frac{z + z_0}{2} \sin \pi \frac{z - z_0}{2}$$

and

$$\cos \pi z - \cos \pi z_0 = -2 \sin \pi \frac{z + z_0}{2} \sin \pi \frac{z - z_0}{2},$$

can be written down immediately according to c):

$$\sin \pi z - \sin \pi z_0 = \pi(z - z_0) \prod_{n=1}^{\infty} \left(1 - \frac{(z + z_0)^2}{(2n-1)^2}\right) \left(1 - \frac{(z - z_0)^2}{(2n)^2}\right)$$

$$\cos \pi z - \cos \pi z_0 = -\tfrac{1}{2} \pi^2 (z^2 - z_0^2) \prod_{n=1}^{\infty} \left(1 - \frac{(z + z_0)^2}{4n^2}\right) \left(1 - \frac{(z - z_0)^2}{4n^2}\right);$$

however, what was said in b) about the form of the product is valid here too.

12. For a suitable choice of the k_ν, the product appearing in K II, § 2, p. 18 again accomplishes what is required. In fact, it is sufficient to choose the k_ν so that the series (3) in K II, p. 17 now converges at least for all $|z| < 1$. This is the case, e.g., for $k_\nu = \nu + \alpha_\nu$. For if z is fixed, $|z| < \rho < 1$, and $\rho < \rho_1 < 1$, then, for all sufficiently large ν, obviously $\left| \alpha_\nu \left(\frac{z}{z_\nu}\right)^{k_\nu} \right| \leqq \alpha_\nu \rho_1^{\alpha_\nu} \cdot \rho_1^\nu$. This is $< K \cdot \rho_1^\nu$, where K denotes a bound for the sequence $n\rho_1^n$, because $n\rho_1^n \to 0$.

The proof of the assertion now proceeds exactly as in K II, § 2. We have to establish the uniform convergence of the series (4) there in $|z| \leqq \rho < 1$, ρ fixed. Minor alterations in the estimates in K II, § 2 — in (5) replace $\frac{1}{2}$ by ρ_1 — show that, if K_1 is a suitable positive number,

$$|f_\nu(z)| \leqq K_1 \cdot \alpha_\nu \left| \frac{\rho}{z_\nu} \right|^{k_\nu}$$

for all $|z| \leqq \rho$ and all sufficiently large ν, which completes the proof.

13. One has merely to choose the z_n so that they cluster along the whole boundary of the unit circle. This can be done, e.g., by taking a circle with center 0 and radius $\left(1 - \frac{1}{k}\right)$, marking on its circumference

the vertices of an inscribed regular k-gon ($k = 3, 4, \ldots$), and arranging these points in a sequence $z_1, z_2, \ldots$.

§ 10. *Entire Functions*

1. According to § 7, Prob. 11, but also already according to the Casorati-Weierstrass theorem itself, the function can only be an entire rational function. By the fundamental theorem of algebra, it must be of the first degree.

2. If $w = g(z)$ is an entire function, and the inverse function $z = g_1(w)$ is also an entire function, then $g(z)$ assumes every value once and only once. Hence, according to Prob. 1, $g(z)$ is linear.

3. The function $g_1(z)$ is again an entire function; in particular, $g_1(z_0) = g'(z_0)$. If, now, invariably $|g(z)| < K$, then, for $|z - z_0| = R$, invariably $|g_1(z)| < \frac{2K}{R}$. By K I, § 20, Theorem 5, this holds also for all $|z - z_0| < R$, and in particular at z_0. Therefore $|g'(z_0)| < \frac{2K}{R}$, and since R may be arbitrarily large, $g'(z_0) = 0$. Since z_0 was arbitrary, $g(z)$ is a constant.

4. Yes. An example is $g(z) = e^z + z$. For if $|\varphi| < \frac{\pi}{2}$ in $z = re^{i\varphi}$, then $|g(z)| \geqq e^{r \cos \varphi} - r$, which, for fixed φ, increases with r to $+\infty$. If, however, $+\frac{\pi}{2} \leqq \varphi \leqq \frac{3\pi}{2}$, then e^z remains bounded, and $|g(z)| \geqq r - e^{r \cos \varphi}$ again tends with r to $+\infty$.

5. Since $f^{(n)}(z_0) \to 0$, the expansion

$$f(z) = \sum_{n=0}^{\infty} \frac{f^{(n)}(z_0)}{n!} (z - z_0)^n$$

converges everywhere, so that $f(z)$ is an entire function. Let $\varepsilon > 0$ be arbitrary, and let k_0 be chosen so that $|f^{(k+1)}(z_0) + \cdots + f^{(k+p)}(z_0)| < \varepsilon$ for $k > k_0$, $p \geqq 1$. Then, for any z_1,

$$|f^{(k+1)}(z_1) + \cdots + f^{(k+p)}(z_1)|$$
$$= \left| \sum_{n=0}^{\infty} \frac{f^{(n+k+1)}(z_0) + \cdots + f^{(n+k+p)}(z_0)}{n!} (z_1 - z_0)^n \right|,$$

and hence $< \varepsilon \cdot e^{|z_1 - z_0|}$ for $k > k_0$, $p \geqq 1$. Therefore $\sum_{n=0}^{\infty} f^{(n)}(z_1)$ converges.

6. Let $W(z)$ be the function, to be constructed according to Weierstrass's factor-theorem, which has a simple zero at each of the z_ν. Then $W'(z_\nu) \neq 0$. Let $M(z)$ be the function, to be constructed according to Mittag-Leffler's theorem, which has a simple pole with the residue $w_\nu / W'(z_\nu)$ at each of the z_ν. Then $g(z) = W(z) \cdot M(z)$ is obviously an entire function which behaves as required.

7. Choose the real rational numbers a_ν in succession so that

1) $|a_0 - \beta| < 1$, 2) $|a_0 + a_1\alpha - \beta| < 1, \cdots$, n) $|a_0 + a_1\alpha + \cdots + a_{n-1}\alpha^{n-1} - \beta| < \dfrac{1}{(n-1)!}, \cdots$. Then, at any rate, $a_0 + a_1\alpha + a_2\alpha^2 + \cdots = \beta$. Further,

$$|a_n\alpha^n| < \frac{1}{n!} + \frac{1}{(n-1)!} = \frac{n+1}{n!},$$

from which it follows that $\sum a_n z^n$ converges everywhere.

8. One can always find a rational complex number arbitrarily close to a given complex number. The preceding proof therefore remains valid without change.

9. If $r' > r$, then the maximum of $|g(z)|$ in $|z| \leqq r'$ cannot be smaller than in $|z| \leqq r$. $M(r)$ thus increases monotonically. If $R > r$, then $g(z)$ is uniformly continuous in $|z| \leqq R$. Therefore, if $\varepsilon > 0$ is given, one can choose $\delta > 0$ so small, that the difference $|g(z') - g(z)| < \varepsilon$ for all r' in $r - \delta < r' < r + \delta$, where $|z| = r$, $|z'| = r'$, and am $z =$ am z'. Consequently, for these r', also $M(r) - \varepsilon < M(r') < M(r) + \varepsilon$. Thus, $M(r)$ is continuous.

10. If, in $g(z) = \sum_{n=0}^{\infty} a_n z^n$, one can choose the point $z = z_0$ on $|z| = r$ so that all the terms $a_n z_0^n$ have the same amplitude (i.e., the terms different from 0), then for this r obviously $M(r) = |g(z_0)|$. Therefore,

a) for e^z: $M(r) = e^r$; b) for sin z: $M(r) = \frac{1}{2}(e^r - e^{-r})$;

c) for cos z: $M(r) = \frac{1}{2}(e^r + e^{-r})$;

d) for $\frac{\sin\sqrt{z}}{\sqrt{z}} = 1 - \frac{z}{3!} + \frac{z^2}{5!} - + \cdots$: $M(r) = \frac{1}{2\sqrt{r}}(e^{\sqrt{r}} - e^{-\sqrt{r}})$.

11. An example is $f(z) = z^2 + 2iz + 1$. For $z = r(\cos\varphi + i\sin\varphi)$ we have

$$|f(z)|^2 = r^4 + 4r^2 + 1 + 2r^2\cos 2\varphi - 4r(1 - r^2)\sin\varphi.$$

For r fixed in $0 < r \leqq \sqrt{2} - 1$, this function of φ reaches a maximum for $\varphi = -\frac{\pi}{2}$; in $\sqrt{2} - 1 \leqq r \leqq 1$, however, for $\sin\varphi = -\frac{1 - r^2}{2r}$. Consequently,

$$M(r) = \begin{cases} 1 + 2r - r^2 & \text{for } 0 \leqq r \leqq \sqrt{2} - 1, \\ (1 + r^2)\sqrt{2} & \text{for } \sqrt{2} - 1 \leqq r \leqq 1. \end{cases}$$

This function is continuous, but not analytic, at the point $r = \sqrt{2} - 1$, because it has a first, but not a second, derivative there.

§ 11. *Partial-fractions Series. Mittag-Leffler's Theorem*

1. a) The desired expansion

$$\pi \tan\frac{\pi z}{2} = \sum_{k=0}^{\infty} \frac{4z}{(2k+1)^2 - z^2}, \quad z = \pm 1, \pm 3, \cdots,$$

follows immediately from the cotangent expansion (K. II, § 6, Ex. 1), since

$$\pi \tan\frac{\pi z}{2} = \pi\cot\frac{\pi z}{2} - 2\pi\cot\pi z.$$

This formula is valid at first only for $z \neq 0, \pm 1, \pm 2, \pm 3, \ldots$, but one can then verify directly that it still holds for $z = 0, \pm 2, \pm 4, \ldots$.

b) Since $\frac{1}{\sin z} = \cot z + \tan\frac{z}{2}$, we have

$$\frac{\pi}{\sin\pi z} = \frac{1}{z} + \sum_{k=1}^{\infty} (-1)^{k-1}\frac{2z}{k^2 - z^2}.$$

c) According to b), upon writing $\frac{2z}{k^2 - z^2} = \frac{1}{k - z} - \frac{1}{k + z}$ and then replacing z by $\frac{1}{2} - z$,

$$\frac{\pi}{\cos\ \pi z} = \frac{2}{1 - 2z} + \left(\frac{2}{1 + 2z} - \frac{2}{3 - 2z}\right)$$
$$- \left(\frac{2}{3 + 2z} - \frac{2}{5 - 2z}\right) + - \cdots .$$

Here we may omit the parentheses and combine pairs of terms having the same sign (why?); this yields

$$\frac{\pi}{\cos\ \pi z} = \frac{4 \cdot 1}{1^2 - 4z^2} - \frac{4 \cdot 3}{3^2 - 4z^2} + \frac{4 \cdot 5}{5^2 - 4z^2} - + \cdots .$$

d) We have $\frac{1}{e^z - 1} = - \frac{1}{2} + \frac{i}{2} \cot \frac{iz}{2}$, and hence

$$= - \frac{1}{2} + \frac{1}{z} + \sum_{k=1}^{\infty} \frac{2z}{z^2 + 4\pi^2 k^2} .$$

e) Since $\cos\ \pi z - \sin \pi z = \sqrt{2} \cdot \sin \pi (\frac{1}{4} - z)$, we obtain, according to b),

$$\frac{\pi}{\cos\ \pi z - \sin\ \pi z} = \frac{1}{\sqrt{2}} \left\{ \frac{1}{\frac{1}{4} - z} + \sum_{k=1}^{\infty} (-1)^{k-1} \frac{2(\frac{1}{4} - z)}{k^2 - (\frac{1}{4} - z)^2} \right\} .$$

2. $M_0(z) = \frac{\alpha_0}{z} + \sum_{\nu=1}^{\infty} \left[\frac{\alpha_\nu}{z - z_\nu} - g_\nu(z) \right]$, where $-g_\nu(z)$ denotes a suitably long initial part of the series $\alpha_\nu \sum_{n=0}^{\infty} \frac{z^n}{z_\nu^{n+1}}$. Now choose $R > 0$ and let it be fixed. Then, according to K II, p. 41, m can be determined so that the terms of the first series for $\nu > m$ are regular in $|z| \leqq R$ and that the series beginning with $\nu = m + 1$ converges uniformly there. One may therefore integrate term by term (say rectilinearly from 0 to z) and place the result in the exponent of e. After including the initial terms, it follows that the product

$$z^{\alpha_0} \cdot \prod_{\nu=1}^{\infty} \left[\left(1 - \frac{z}{z_\nu}\right) e^{G_\nu(z)} \right]^{\alpha_\nu},$$

where $G_\nu(z)$ is of the form $\frac{z}{z_\nu} + \frac{1}{2}\left(\frac{z}{z_\nu}\right)^2 + \cdots + \frac{1}{n_\nu}\left(\frac{z}{z_\nu}\right)^{n_\nu}$, represents an entire function with the properties required in Weierstrass's theorem.

3. The theorem and its proof are not altered in the least if (all or some of) the points z_ν are permitted to be essential singular points instead of poles.

4. If, in addition to the descending part, the entire ascending part is also prescribed, then the function possessing this expansion at z_ν is thereby completely determined. Therefore no further conditions can be imposed on it. An (arbitrarily long) initial section of the ascending part can, however, be prescribed at *every* point z_ν. Thus, for $\nu = 0, 1, 2, \ldots$, let

$$H_\nu(z) = \sum_{n=-\infty}^{\beta_\nu} a_n^{(\nu)} (z - z_\nu)^n,$$

where $\beta_0, \beta_1, \ldots$ denote any nonnegative integers. Then there exists a single-valued function $F(z)$ which is regular in the whole plane except at the points z_ν, and whose behavior at each of the points z_ν is such that $F(z) - H_\nu(z)$ is regular there and possesses a zero whose order is at least $\beta_\nu + 1$. The proof can be effected by continuing the idea in § 10, Prob. 6. Let $W(z)$ be an entire function which has a zero of order $\beta_\nu + 1$ at z_ν for every ν, but which $\neq 0$ otherwise. Let $h_\nu(z)$ be the principal part of the Laurent expansion of $H_\nu(z)/W(z)$ at the point z_ν. Now form a function $M(z)$ which has the principal parts $h_\nu(z)$ at the points z_ν but which is single-valued and regular otherwise. Then $F(z) = W(z) \cdot M(z)$ fulfills the requirements. For in a neighborhood of z_ν,

$$M(z) = \frac{H_\nu(z)}{W(z)} + f_\nu(z), \qquad (f_\nu(z) \text{ regular at } z_\nu).$$

Thus $W(z) \cdot M(z) - H_\nu(z) = W(z) \cdot f_\nu(z)$ has a zero at z_ν of order not less than $\beta_\nu + 1$.

5. The proof is quite analogous to that in K II, § 5. The expansion $h_\nu(z) = a_0^{(\nu)} + a_1^{(\nu)} z + \cdots$ $(\nu = 1, 2, \ldots)$ is uniformly convergent now

for $|z| \leqq |z_\nu|^2$, and $g_\nu(z)$ can be chosen so that $|h_\nu(z) - g_\nu(z)| < \frac{1}{2^\nu}$ for $|z| \leqq |z_\nu|^2$. Then $M_0(z) = h_0(z) + \sum_{\nu=1}^{\infty} [h_\nu(z) - g_\nu(z)]$ again satisfies the requirements. For if ρ is given in $0 < \rho < 1$, then m can be chosen so that $|z_\nu|^2 > \rho$ for $\nu > m$. Therefore $\sum_{\nu=m+1}^{\infty} [h_\nu(z) - g_\nu(z)]$ is uniformly convergent in $|z| \leqq \rho$ and has terms which are regular in $|z| \leqq \rho$, so that it represents a regular function there. Consequently $M_0(z)$ possesses the required properties in $|z| < \rho$ and, since ρ was arbitrary, in $|z| < 1$.

6. The proof is based on the same idea as that used in the preceding solution. We carry it out only for the case in which ∞ belongs neither to $\mathfrak{M}$ nor to $\mathfrak{M}'$. About every point z_ν we describe the smallest circle — call its radius ρ_ν — whose boundary contains at least one point z'_ν of $\mathfrak{M}'$ but which has no point of $\mathfrak{M}'$ in its interior. Since $\mathfrak{M}'$ is closed (why?), such a circle invariably exists, and $\rho_\nu \to 0$. Let the expansion of $h_\nu(z)$ for the exterior region $|z - z'_\nu| > \rho_\nu$ be $h_\nu(z) = b_0^{(\nu)} + \dfrac{b_1^{(\nu)}}{z - z'_\nu} + \cdots$. Since it converges uniformly for $|z - z'_\nu| \geqq 2\rho_\nu$, one can choose such a long initial part of the series [denote it again by $g_\nu(z)$] that $|h_\nu - g_\nu| < \frac{1}{2^\nu}$ for $|z - z'_\nu| \geqq 2\rho_\nu$. Then

$$F(z) = \sum_{\nu=1}^{\infty} [h_\nu(z) - g_\nu(z)]$$

possesses the required properties. For let $\overline{\mathfrak{G}}$ be a closed region which (including its boundary) contains no point of $\mathfrak{M}'$. Then the smallest distance between a point of $\overline{\mathfrak{G}}$ and a point of $\mathfrak{M}'$ is still positive. Call it p. We now choose m so large that $\rho_\nu < \frac{1}{2}p$ for $\nu > m$. Then the terms of the series $F_m(z) = \sum_{\nu=m+1}^{\infty} (h_\nu - g_\nu)$ are regular in $\overline{\mathfrak{G}}$ and $< 1/2^\nu$ in absolute value. Hence, F_m is regular in $\overline{\mathfrak{G}}$, and F fulfills the requirements in $\overline{\mathfrak{G}}$. Since $\overline{\mathfrak{G}}$ was arbitrary, F satisfies the conditions of the problem.

Remark: If ∞ belongs to $\mathfrak{M}$ or to $\mathfrak{M}'$, one has to introduce a point ζ

which belongs neither to $\mathfrak{M}$ nor to $\mathfrak{M}'$, and expand the $h_\nu(z)$ in powers of $\left(\frac{z-\zeta}{z-z_\nu'}\right)$.

7. The theorem reads: Let $\mathfrak{M}$ be an arbitrary isolated point set whose points are $z_1, z_2, \ldots$, let $\mathfrak{M}'$ be the set of limit points, and associate with every z_ν a natural number α_ν. Then one can set up an infinite product which converges uniformly in every closed region $\overline{\mathfrak{G}}$ that contains no point of $\mathfrak{M}'$, and which in $\overline{\mathfrak{G}}$ represents a regular function having a zero of order α_ν at every z_ν belonging to $\overline{\mathfrak{G}}$, but which $\neq 0$ otherwise in $\overline{\mathfrak{G}}$. The proof is effected, exactly as in the case of Prob. 2, by applying the theorem of the preceding problem with $h_\nu(z) = \frac{\alpha_\nu}{z - z_\nu}$ and placing the integral of the constructed function, taken from a fixed point z_0 in $\overline{\mathfrak{G}}$ to z, in the exponent of e. One obtains

$$W(z) = \prod_{\nu=1}^{\infty} \left[\left(1 - \frac{z_\nu - z_\nu'}{z - z_\nu}\right) e^{G_\nu(z)}\right]^{\alpha_\nu},$$

where $G_\nu(z)$ is of the form

$$\left(\frac{z_\nu - z_\nu'}{z - z_\nu}\right) + \frac{1}{2}\left(\frac{z_\nu - z_\nu'}{z - z_\nu}\right)^2 + \cdots + \frac{1}{n_\nu}\left(\frac{z_\nu - z_\nu'}{z - z_\nu}\right)^{n_\nu}.$$

8. The solution is obtained exactly as that of § 10, Prob. 6 by applying the generalizations of Weierstrass's factor-theorem and Mittag-Leffler's partial-fractions-theorem given in the preceding two problems.

9. The solution is obtained exactly as that of Prob. 4 by applying the generalizations of Weierstrass's factor-theorem and Mittag-Leffler's partial-fractions theorem given in Probs. 6 and 7.

§ 12. *Meromorphic Functions*

1. We show first that the sequence of functions

$$\frac{1}{g_n(z)} = \tilde{g}_n(z) = \frac{z(z+1)\cdots(z+n)}{n^z n!}, \quad n = 1, 2, \cdots,$$

converges uniformly in $|z| \leqq R$ for every $R > 0$. In fact (see K II, p. 32), $\tilde{g}_n(z) = e^{\delta_n z} \cdot P_n(z)$, if we set

$$\left(1 + \tfrac{1}{2} + \cdots + \tfrac{1}{n} - \log n\right) = \delta_n, \quad z \cdot \prod_{\nu=1}^{n} \left[\left(1 + \frac{z}{\nu}\right) e^{-\frac{z}{\nu}}\right] = P_n(z).$$

According to K II, pp. 11–14 and the proof of Weierstrass's factor-theorem, the sequence of $P_n(z)$ is uniformly convergent in $|z| \leqq R$. Since $\delta_n \to C$, this is obviously also true of the sequence $e^{\delta_n z}$. From this one easily infers the assertion.

We proceed, however, as follows: From what we have said, there exists at any rate a constant A such that for all n and all $|z| \leqq R$ invariably $|\tilde{g}_n(z)| < A$. But then for all these n and z (cf. § 2, Prob. 4),

$$|\tilde{g}_{n+1}(z) - \tilde{g}_n(z)| \leqq A \cdot \left| \left(1 + \frac{z}{n+1}\right)\left(1 - \frac{1}{n+1}\right)^z - 1 \right| < \frac{B}{n^2},$$

where B denotes a suitable new constant. Therefore $\sum_{n=1}^{\infty} (\tilde{g}_{n+1} - \tilde{g}_n)$, and in fact $\sum_{n=1}^{\infty} |\tilde{g}_{n+1} - \tilde{g}_n|$, converges uniformly in $|z| \leqq R$. The same is therefore true of $\lim_{n\to\infty} \tilde{g}_n(z)$.

Now let $\mathfrak{G}$ be a region of the sort mentioned in the problem. Then all terms of the sequence $\tilde{g}_n(z)$ are different from 0 in $\mathfrak{G}$, and the same holds for the limit. Consequently, there exists a positive number α such that, for all z in $\mathfrak{G}$ and all n, invariably $|\tilde{g}_n(z)| \geqq \alpha$. But then $\dfrac{1}{\tilde{g}_n(z)} = g_n(z)$, together with $\tilde{g}_n(z)$, is also uniformly convergent in $\mathfrak{G}$, as is evident from the estimate $|g_{n+k} - g_{n+1}| \leqq \dfrac{1}{\alpha^2} |\tilde{g}_{n+k} - \tilde{g}_{n+1}|$ in $\mathfrak{G}$.

The estimate $|g_{n+1} - g_n| \leqq \dfrac{1}{\alpha^2} |\tilde{g}_{n+1} - \tilde{g}_n|$ shows at the same time that the series $\sum_{n=1}^{\infty} |g_{n+1} - g_n|$ also converges uniformly in $\mathfrak{G}$.

2. Let $g_n(z)$ have the same meaning as in Prob. 1. Then the limit in question is $\lim_{n\to\infty} \dfrac{g_n(z_2)}{g_n(z_1)} \cdot n^{z_1 - z_2}$. This limit exists, and is then $= 0$, if, and

only if, $\Re(z_2) > \Re(z_1)$, because $g_n(z)$ has a finite limit different from 0 for every z under consideration.

3. Let $f(z)$ be convergent for a definite $z \neq 0, -1, -2, \ldots$. Then, if $g_n(z)$ has the same meaning as in Prob. 1,

$$\frac{n! a_n}{z(z+1)\cdots(z+n)} = \frac{a_n}{n^z} \cdot g_n(z).$$

According to I, § 3, Prob. 12, the series $F(z)$ therefore is also convergent. For the conditions of that problem are certainly fulfilled if the series Σa_n and $\Sigma |b_n - b_{n+1}|$ appearing there converge. The convergence of the latter in our case (i.e., for $b_n = g_n(z)$), however, was shown in connection with Prob. 1 of this paragraph.

The converse follows in exactly the same manner from $\frac{a_n}{n^z} = \frac{n!\, a_n}{z(z+1)\cdots(z+n)} \cdot \tilde{g}_n(z)$, since $\Sigma |\tilde{g}_n - \tilde{g}_{n+1}|$ was also shown to be convergent.

4. This follows just as in the preceding problem, on applying I, § 9, Prob. 4. For, according to Prob. 1, the two series $\Sigma |g_{n+1}(z) - g_n(z)|$ and $\Sigma |\tilde{g}_{n+1}(z) - \tilde{g}_n(z)|$ are uniformly convergent in every region of the sort now coming under consideration.

5. This follows exactly as in Prob. 3, since the convergence of $\Sigma |g_n - g_{n+1}|$ implies that of $\Sigma \big| |g_n| - |g_{n+1}| \big|$ — and analogously for $\tilde{g}_n$.

6. Let $g_n(z)$ have the same meaning as in Prob. 1. Then we are dealing with the limit of

$$\frac{g_n(z)}{g_{n+p_n}(z)} \cdot \left(1 + \frac{p_n}{n}\right)^z.$$

Here the first factor $\to 1$ (obviously also for $z = 0, -1, -2, \ldots$ with suitable interpretation). Therefore the limit of the whole expression $= 1$ if, and only if, $p_n/n \to 0$.

7. From the factor representation of $1/\Gamma(z) = K(z)$ in K II, § 3, 3 we obtain by logarithmic differentiation

$$\frac{\Gamma'(z)}{\Gamma(z)} = -C - \frac{1}{z} + \sum_{\nu=1}^{\infty} \frac{z}{\nu(z+\nu)}.$$

Now (draw a figure) for all z and ν under consideration,

1) $|z| \geqq 2$, 2) $|z+\nu| \geqq |z+1|$, 3) $|z+\nu| \geqq \nu - 1$.

It follows that

$$\left|\frac{\Gamma'(z)}{\Gamma(z)}\right| \leqq 1 + \tfrac{1}{2} + |z| \cdot \sum_{\nu=1}^{[|z|]} \frac{1}{\nu \cdot |z+1|} + |z| \cdot \sum_{\nu=[|z|]+1}^{\infty} \frac{1}{\nu(\nu+1)}$$

$$< 2 + \frac{|z|}{|z+1|} (\log|z| + 1) + |z| \cdot \frac{1}{|z| - 1} < A \cdot \log|z|.$$

8. According to I, § 10, Prob. 3a, $b_n = \frac{1}{n!} \sum_{k=2}^{\infty} \frac{(\log k)^n}{k^2}$. We are thus concerned with proving the relation

$$(*) \qquad \sum_{k=2}^{\infty} \frac{(\log k)^n}{k^2} \cong n!$$

Differentiation shows that $\frac{(\log x)^n}{x^2}$ increases monotonically until $x = e^{\frac{n}{2}}$ where it attains a maximum and then decreases monotonically. The maximum itself is equal to $\frac{1}{2^n}\left(\frac{n}{e}\right)^n$, which is $< \frac{1}{2^n} n!$ because $n^n e^{-n} < n!$ [1])

Let us set $\left[e^{\frac{n}{2}}\right] + 1 = m$. Then, for $k \geqq m$,

$$\frac{\log^n (k+1)}{(k+1)^2} < \int_k^{k+1} \frac{\log^n x}{x^2}\, dx < \frac{\log^n k}{k^2}.$$

Write this down for $k = m, m+1, \ldots$ and add. If we denote the sum of our series in (*) by S_n, it follows that

$$S_n - A_m < \int_1^{\infty} \frac{\log^n x}{x^2}\, dx - J_m < S_n - A_{m-1};$$

here A_p denotes the pth partial sum of our series and J_m denotes the integral from 1 to m. Now $(x = e^t)$

[1]) This is obtained by multiplying together the familiar inequalities $\left(1 + \frac{1}{\nu}\right)^{\nu} < e$ for $\nu = 1, 2, \ldots, n$.

$$\int_1^\infty \frac{\log^n x}{x^2}\, dx = n!$$

and consequently

$$\frac{J_m - A_m}{n!} < 1 - \frac{1}{n!}\, S_n < \frac{J_m - A_{m-1}}{n!}.$$

The absolute value of the left-hand side, as well as that of the right-hand side, is, because of the maximum calculated above,

$$< \frac{2}{n!} \cdot \frac{n!}{2^n} \cdot \left(e^{\frac{n}{2}} + 1\right) < 4 \left(\frac{\sqrt{e}}{2}\right)^n.$$

Since this $\to 0$, our assertion $S_n/n! \to 1$ is established. It shows that $+1$ is a singular point of the ζ-function. Our estimate shows further, however, that $\zeta(z) - \sum_{n=0}^{\infty} (-1)^n\ (z-2)^n = \zeta(z) - \frac{1}{z-1}$ is regular at least in the circle $|z-2| < \frac{2}{\sqrt{e}}$, so that $\zeta(z)$ has only a pole of the first order with the residue $+1$ at the point $+1$.

Cf. in this connection the remark at the end of the solution of the next problem.

9. If we make an estimate of those terms of the series in (*) (see the preceding problem) for which $k \leqq m$, analogous to that which we just carried out for the terms with $k \geqq m$, we find more precisely that S_n and the integral $\int_1^\infty \frac{\log^n x}{x^2}\, dx = n!$ differ by an amount which is less than two terms of the series, i.e., that

$$|\, S_n - n!\, | < 2\, \frac{n!}{2^n}, \quad \left| \frac{1}{n!}\, S_n - 1 \right| < \frac{2}{2^n}.$$

This shows that $\zeta(z)$ is actually regular in $|z-2| < 2$, except at the pole $+1$ determined above.

The sum of the series in (*) which appears in the solution of the preceding problem can be estimated in a substantially more accurate manner with the aid of what is known as Euler's summation formula, and thereby one can show directly, in a somewhat different fashion than in K II, § 6, Ex. 4, that $\zeta(z) - \frac{1}{z-1}$ is an entire function.

CHAPTER IV

PERIODIC FUNCTIONS

§ 13. *Simply Periodic Functions*

1. For otherwise it would also have all numbers $n + n'\sqrt{2}$ ($n, n' = 0, \pm 1, \pm 2, \ldots$) as periods. These numbers, however, lie everywhere dense on the axis of reals, i.e., every point there is a limit point of the numbers mentioned. According to K II, § 7, Theorem 2, this cannot occur.

2. A periodic function assumes every value that it assumes at all, infinitely often; a rational function, however, does not.

Preliminary remarks to Solutions 3 *to* 7: Problems 3 to 7 are resolved very simply by the remark that, in the transformation of the functions $f(z)$ into the rational functions $\varphi(\zeta)$ carried out in K II, p. 66 et seq., to the "upper end" of the period strip corresponds obviously, by virtue of $\zeta = e^{2\pi i z}$, an approach to $\zeta = 0$, and likewise to the "lower end" corresponds an approach to $\zeta = \infty$.

3. (Cf. the preliminary remark.) If $f(z)$ is single-valued and regular in the strip, then $\varphi(\zeta)$ is single-valued and regular in the entire plane except perhaps at 0 or ∞. If, in addition, $\varphi(\zeta)$ remained bounded as ζ approached 0 and as ζ approached ∞, then, according to Riemann's theorem, K I, § 32, Theorem 3, $\varphi(\zeta)$ would also be regular at 0 and ∞. A nonconstant function, however, cannot be regular in the whole plane including ∞.

4. (Cf. the preliminary remark to Prob. 3, and the solution of Prob. 3). If $\varphi(\zeta)$ remains bounded as $\zeta \to 0$, then, according to Riemann's theorem, $\varphi(\zeta)$ is regular at $\zeta = 0$, so that $\lim \varphi(\zeta)$ exists as $\zeta \to 0$. Therefore $f(z)$ has a limit as z tends toward the upper end of the strip. An analogous treatment applies to the lower end. The order of the zero of $\varphi(\zeta) - a$ at $0, \infty$, respectively, is taken as the order to which $f(z)$ assumes the value a.

5. (Cf. the preliminary remark to Prob. 3, and the preceding solutions.) If $\varphi(\zeta)$ does not remain bounded as $\zeta \to 0$, then $\varphi(\zeta)$, being a rational function, must become definitely infinite (i.e., $|\varphi(\zeta)| > G$ for $|\zeta| < \delta$) and must therefore have a pole at $\zeta = 0$. The order of this pole is taken as the order of the pole of $f(z)$ at the upper end of the strip. The lower end is handled similarly.

6. (Cf. the preliminary remark to Prob. 3, and the preceding solutions.) For, a rational function assumes every value (including ∞) equally often, provided that an a-point of order α is, as usual, counted as an a-point α times. (A pole is regarded here as an ∞-point.)

7. (Cf. the preliminary remark to Prob. 3, and the preceding solutions.) For $\varphi(\zeta)$ is supposed to be a rational function (see K I, § 35, Theorems 1 and 2).

§ 14. *Doubly Periodic Functions*

1. All periods have the form $n\omega + n'\omega$. For two periods, say $\tilde{\omega} = k\omega + k'\omega'$ and $\tilde{\omega}' = l\omega + l'\omega'$, to form another pair of primitive periods (k, l, n, k', ... here denote real integers), it is necessary and sufficient that all periods be representable also in the form $n\tilde{\omega} + n'\tilde{\omega}'$. From the definition of $\tilde{\omega}$ and $\tilde{\omega}'$ it now follows that

$$\omega = \frac{1}{kl' - k'l}(l'\tilde{\omega} - k'\tilde{\omega}'), \quad \omega' = \frac{1}{kl' - k'l}(-l\tilde{\omega} + k\tilde{\omega}').$$

For the coefficients of $\tilde{\omega}$ and $\tilde{\omega}'$ to be integers too, it is necessary that $kl' - k'l = \delta$ be a divisor of k, k', l, l', and hence that δ^2 be a divisor of δ, δ itself be a divisor of 1, and consequently $\delta = \pm 1$. This is also sufficient; for if $\delta = \pm 1$, then ω and ω', and therefore all periods, can be represented in the form $n\tilde{\omega} + n'\tilde{\omega}'$. Thus, one obtains all pairs of primitive periods $\tilde{\omega}$, $\tilde{\omega}'$, if k, k', l, l' run through all integers for which $kl' - k'l = \pm 1$ (geometrical meaning?).

2. The difference in question is equal to the following integral taken along the boundary of a suitable period-parallelogram (cf. Prob. 5 in § 8 and the proof of K II, § 9, Theorem 4):

$$\frac{1}{2\pi i}\int z\frac{f'(z)}{f(z)}\,dz = \frac{1}{2\pi i}\cdot\left\{\int_a^{a+\omega} + \int_{a+\omega}^{a+\omega+\omega'} + \int_{a+\omega+\omega'}^{a+\omega'} + \int_{a+\omega'}^{a}\right\}.$$

Let us introduce $z' = z - \omega'$ as a new variable in the third integral. Combining this integral with the first we obtain the value

$$-\,\omega'\int_a^{a+\omega}\frac{f'(z)}{f(z)}\,dz = -\,\omega'\cdot\Big[\log f(z)\Big]_a^{a+\omega} = 2k'\pi i\cdot\omega';$$

the last equality results from the fact that $f(z)$ has the same value at a and $a + \omega$ so that the logarithms differ only by a multiple of $2\pi i$. Similarly the second and fourth integrals combined yield the value $2k\pi i\omega$, so that our difference $= k\omega + k'\omega'$.

3. From $f(\tilde{\omega} - z) = f(-z) = -f(z)$ it follows, for $z = \frac{1}{2}\tilde{\omega}$, that $f(\frac{1}{2}\tilde{\omega}) = -f(\frac{1}{2}\tilde{\omega})$, which is only possible if $f(\frac{1}{2}\tilde{\omega}) = 0$ or $= \infty$. Since $f(z)$ also proves to be an odd function of $z' = z - \frac{1}{2}\tilde{\omega}$, the Laurent expansion of $f(z)$ at the point $\frac{1}{2}\tilde{\omega}$ has only odd powers.

4. a) The fundamental parallelogram is first mapped by means of $z' = \dfrac{2\pi i}{\omega}z$ on that parallelogram which is spanned by the vectors extending from 0 to $2\pi i$ and to $2\pi i\frac{\omega'}{\omega}$. This parallelogram is then mapped by means of $w = e^{z'}$ on the ring $1 \geqq |w| \geqq e^{-2\pi\tau_1}$, where τ_1 denotes the imaginary part of ω'/ω and may be taken to be positive according to K II, p. 63, footnote 2. b) To the translation (ω') corresponds a stretching in the ratio $1 : e^{-2\pi\tau_1}$ in the w-plane. Thus, to the successive parallelograms correspond rings whose radii form a geometric progression and which fill out the punctured plane $0 < |w| < +\infty$ precisely once.

5. If $f(z) = \varphi(\zeta)$, then also $f(z + \omega) = \varphi(\zeta)$, but $f(z + \omega') = \varphi(\mu\zeta)$ with $\mu = e^{2\pi i\frac{\omega'}{\omega}}$. Thus, if $f(z)$ has the period ω', then $\varphi(\mu\zeta) = \varphi(\zeta)$, i.e., $\varphi(\zeta)$ has a "multiplicative period".

6. $e^{\wp(z)}$ is doubly periodic, but not elliptic. For, the lattice points of the network of periods are essential singular points of the function,

whereas an elliptic function has only poles in the finite part of the plane.

7. We set $\omega = 2\alpha > 0$ and $-i\omega' = 2\beta > 0$. In the expansion (see K II, p. 46)

$$\wp(z) = \frac{1}{z^2} + \sum_{k,k'}{}' \left[\frac{1}{(z - 2k\alpha - 2k'i\beta)^2} - \frac{1}{(2k\alpha + 2k'i\beta)^2}\right],$$

we name the terms after the lattice points (k, k') in a kk'-coordinate system; and for a fixed z we designate the two terms in the brackets by $(k, k')_1$, $(k, k')_2$, respectively, both together by $[k, k']$. We agree to set $(0, 0)_2 = 0$, $[0, 0] = \frac{1}{z^2}$. Now $(k, k')_2$ and $(k, -k')_2$ are obviously complex conjugates (or both real if $k = 0$ or $k' = 0$).

a) If $z = x > 0$ is real, then $(k, k')_1$ and $(k, -k')_1$, and consequently also $[k, k']$ and $[k, -k']$, are complex conjugates for $k' \gtrless 0$, real for $k' = 0$. The terms of our series are thus real for $k' = 0$, and occur in complex conjugate pairs for $k' \gtrless 0$. Their sum, $\wp(x)$, is therefore real.

b) If $z = \alpha + iy$, then $(k, k')_1 = (\alpha + iy - 2k\alpha - 2k'i\beta)^{-2} = (\alpha - iy + 2(k - 1)\alpha + 2k'i\beta)^{-2}$ is the conjugate of $(\alpha + iy + 2(k - 1)\alpha - 2k'i\beta)^{-2} = (-k + 1, k')_1$. The four terms

$$[k, k'] + [-k + 1, k'] + [k, -k'] + [-k + 1, -k']$$

(coincident ones counted only once) thus yield a real sum. Therefore $\wp(\alpha + iy)$ is also real.

c) If $z = x + i\beta$, then one can show in a similar manner that the four terms

$$[k, k'] + [k, -k' + 1] + [k, -k'] + [k, k' - 1]$$

invariably yield a real sum. Therefore $\wp(x + i\beta)$ is also real.

d) Finally, if $z = iy$, then $(k, k')_1 = (iy - 2k\alpha - 2k'i\beta)^{-2} = (-iy - 2(-k)\alpha + 2k'i\beta)^{-2}$ is the conjugate of $(iy - 2(-k)\alpha - 2k'i\beta)^{-2} = (-k, k')_1$, and we see that now the four terms

$$[k, k'] + [-k, k'] + [k, -k'] + [-k, -k']$$

(coincident ones again counted only once) invariably yield a real sum. Hence, $\wp(iy)$ is also real.

Thus if z, beginning at 0, traverses in the positive sense the quarter $(0 \ldots \alpha \ldots \alpha + i\beta \ldots i\beta \ldots 0)$ of the fundamental parallelogram lying at 0 until it returns to 0, then $\wp(z)$ is invariably real, and in fact (because of the term z^{-2}) is first large and positive and finally large and negative. Hence $w = \wp(z)$ describes the axis of reals from right to left. Each of these real values is assumed only once. For, the values assumed in $0 \ldots \alpha$ are also assumed (in reverse order) on the segment $\alpha \ldots 2\alpha (= \omega)$ because $\wp(z) = \wp(\omega - z)$ — and analogously for the remaining three segments. Every value w, however, is assumed only precisely twice by $\wp(z)$ (see K II, § 9, Theorem 7). (Every value is assumed, moreover, at two distinct points, each to the first order; only the values $\wp(\alpha)$, $\wp(\alpha + i\beta)$, $\wp(i\beta)$ and ∞ are assumed at exactly one point, but then, to make up for it, to the second order.) Consequently (cf. Probs. 3 and 4 in § 20, and B, § 9) the quarter of the rectangle under consideration is mapped in a one-to-one manner on the lower (why *lower*?) half of the w-plane. It follows, finally, by reflection (B, § 10), that the entire fundamental rectangle is mapped on a two-sheeted w-plane which hangs together at the points $\wp(\alpha)$, $\wp(\alpha + i\beta)$, $\wp(i\beta)$, and ∞. (Cf. B, § 14 and § 15, 8.)

8. a) See K II, p. 83, footnote 1. b) According to K II, p. 83, top, and according to a),

$$\wp(z) = \frac{1}{z^2} + 3s_4 z^2 + 5s_6 z^4 + 7s_8 z^6 + \cdots,$$

and hence

$$\wp''(z) = \frac{6}{z^4} + 3\cdot 2\cdot 1 s_4 + 5\cdot 4\cdot 3 s_6 z^2 + 7\cdot 6\cdot 5 s_8 z^4 + \cdots.$$

On the other hand, it follows from $\wp'^2 = 4\wp^3 - 60 s_4 \wp - 140 s_6$ (see K II, p. 83), by differentiation and division by $2\wp'$, that $\wp'' = 6\wp^2 - 30 s_4$, and hence

$$\wp''(z) = 6\left[\frac{1}{z^4} + s_4 + 10 s_6 z^2 + (9s_4^2 + 14 s_8) z^4 + \cdots\right].$$

Consequently, $9s_4^2 + 14s_8 = 35s_8$ or $7s_8 = 3s_4^2$. — By comparing coefficients of higher powers, all further s_{2m} can be expressed as entire rational functions of s_4 and s_6, and hence, of g_2 and g_3, with rational coefficients.

9. $\wp'(z)$ is an odd elliptic function, and therefore (see Prob. 3) vanishes for the half-periods $\frac{1}{2}\omega$, $\frac{1}{2}\omega'$, and $\frac{1}{2}(\omega + \omega')$. At any rate then, $w = \wp(\frac{1}{2}\omega)$, $\wp(\frac{1}{2}\omega')$, $\wp(\frac{1}{2}(\omega + \omega'))$ are roots of $4w^3 - g_2 w - g_3$. The three values mentioned are also distinct, because each of the values is assumed to the second order (since the corresponding derivative vanishes), and no value is assumed more than twice.

10. Let the zeros $z_1, \ldots, z_k$ and the poles $\zeta_1, \ldots, \zeta_k$ of $f(z)$ be denoted as in Prob. 2. Then $\Sigma z_\varkappa - \Sigma \zeta_\varkappa$ is equal to a period ω. Replace ζ_1 by the congruent point $\zeta_1 + \tilde{\omega}$, which possibly lies outside the period-parallelogram first chosen, and denote this point again by ζ_1. Then $\Sigma z_\varkappa = \Sigma \zeta_\varkappa$. According to K II, § 9, p. 87(2), the σ-quotient written down in the problem proves to be doubly periodic with the periods ω and ω'. When divided by $f(z)$ it becomes a doubly periodic entire function, and hence a constant.

CHAPTER V

ANALYTIC CONTINUATION

§ 15. *Behavior of Power Series on the Boundary of the Circle of Convergence*

1. According to I, § 11, Prob. 7, $h(z) \to +\infty$ as $z \to +1$ radially. Now if we set $a_n = g \cdot b_n + \varepsilon_n b_n$, then $\varepsilon_n \to 0$; therefore $\sum a_n z^n$ has a radius at least as large as that of $\sum b_n z^n$. Let $\varepsilon > 0$ be given, and choose m so that $|\varepsilon_n| < \frac{1}{2}\varepsilon$ for $n > m$. Then, for $z = x$ in $0 < x < 1$,

$$\left|\frac{f(x)}{g(x)} - g\right| < \frac{|\varepsilon_0 b_0| + \cdots + |\varepsilon_m b_m|}{h(x)} + \frac{\varepsilon}{2}.$$

By the preliminary remark, $\delta > 0$ can now be assigned so that the right-hand side is $< \varepsilon$ in $1 - \delta < x < 1$.

2. It is no longer true in general. For example, if $h(z) = e^{\left(\frac{1}{1-z}\right)^2} = \sum_{n=0}^{\infty} b_n z^n$, then $b_n > 0$, $\sum b_n$ diverges (why?), but, according to § 3, Prob. 7a, $h(z)$ does not $\to \infty$, but tends to 0 instead, if, e.g., $z \to +1$ in such a manner that am $(1 - z) = \frac{2\pi}{3}$. It is easy to show now that $f(z) = h(z) + \frac{1}{1-z} = \sum a_n z^n = \sum (b_n + 1) z^n$ is a counterexample to the projected extension of the theorem of the preceding problem. If, however, $h(z)$ has the property that, having chosen the triangle $z_1 z_2 1$, there exists a constant $\gamma > 0$ such that $|h(z)|/h(|z|) \geqq \gamma$ for all points z of this triangle different from $+1$, then the theorem remains valid. Indeed, for the z mentioned we now have

$$\left|\frac{f(z)}{h(z)} - g\right| \leqq \left\{\frac{|\varepsilon_0 b_0| + \cdots + |\varepsilon_m b_m|}{h(|z|)} + \frac{\varepsilon}{2}\right\} \cdot \frac{h(|z|)}{|h(z)|}$$

$$\leqq \frac{1}{\gamma} \cdot \varepsilon$$

for all z of the triangle for which $1 - \delta < |z| < 1$, $(\delta = \delta(\varepsilon))$. Since γ is fixed and $\varepsilon > 0$ is arbitrary, this contains the proof.

3. According to I, § 1, Prob. 13, $h(z) = \dfrac{1}{1 - z} = \Sigma z^n$ satisfies the condition $|h(z)|/h(|z|) \geqq \gamma$ formulated in the preceding solution. Hence, as z tends to $+1$ within a fixed triangle $z_1 z_2 1$,

$$\frac{f(z)}{h(z)} = \left(\sum_{n=0}^{\infty} a_n z^n\right) \to \lim_{n\to\infty} \frac{s_n}{1} = \sum_{n=0}^{\infty} a_n.$$

4. From $F(z) = \sum_{n=0}^{\infty} a_n z^n$ we obtain $\dfrac{1}{1 - z} F(z) = \sum_{n=0}^{\infty} s_n z^n$ and $\dfrac{1}{(1 - z)^2} F(z) = f(z) = \sum_{n=0}^{\infty} (s_0 + s_1 + \cdots + s_n) z^n$. We now apply the theorem in Prob. 2 to this function $f(z)$ and to $h(z) = \left(\dfrac{1}{1 - z}\right)^2$: as z approaches 1 within the triangle $z_1 z_2 1$,

$$\frac{f(z)}{h(z)} = F(z) \to \lim_{n\to\infty} \frac{s_0 + s_1 + \cdots + s_n}{n + 1} = s.$$

For we have $h(z) = \sum_{n=0}^{\infty} (n + 1) z^n$, and this $h(z)$, according to I, § 1, Prob. 13, satisfies the condition $h(|z|)/|h(z)| \leqq K^2$ for all $z \neq +1$ of the triangle.

5. a) Proof according to Prob. 4. We have $s_n = 1$ for $(2m - 2)^2 \leqq n < (2m - 1)^2$ and $s_n = 0$ for $(2m - 1)^2 \leqq n < (2m)^2$, $m = 1, 2, \ldots$. Hence, as one can easily verify, $\dfrac{s_0 + s_1 + \cdots + s_n}{n + 1} \to \frac{1}{2}$, which completes the proof.

b) If we multiply and divide by $(1 - z)^{-1}$, the function can be written as follows:

$$\frac{\sum_{n=0}^{\infty} \left([\sqrt{n}] + 1\right) z^n}{(1 - z)^{-3/2}} = \frac{f(z)}{h(z)} = \frac{\Sigma a_n z^n}{\Sigma b_n z^n}.$$

According to Probs. 1 and 2, it suffices to prove $b_n > 0$, the divergence of Σb_n, $\frac{a_n}{b_n} \to \frac{1}{2}\sqrt{\pi}$, and $\frac{|h(z)|}{h(|z|)} \geqq \gamma$. Now certainly $b_n > 0$ and

$$\frac{1}{b_n} = \frac{n!n^{3/2}}{\frac{3}{2}(\frac{3}{2}+1)\cdots(\frac{3}{2}+n)} \cdot \frac{\frac{3}{2}+n}{n^{3/2}} \cong \frac{1}{\sqrt{n}} \cdot \Gamma(\tfrac{3}{2}),$$

so that Σb_n diverges and $\frac{a_n}{b_n} \to \Gamma(\frac{3}{2}) = \frac{1}{2}\Gamma(\frac{1}{2}) = \frac{1}{2}\sqrt{\pi}$. Moreover, according to I, § 1, Prob. 13, $\frac{h(|z|)}{|h(z)|} = \left(\frac{|1-z|}{1-|z|}\right)^{3/2}$ is bounded.

c) Multiplying and dividing again by $(1-z)^{-1}$, we deal with

$$\frac{a_1 z + a_2 z^2 + \cdots + a_n z^n + \cdots}{z + (1+\frac{1}{2}) z^2 + \cdots + \left(1 + \frac{1}{2} + \cdots + \frac{1}{n}\right) z^n + \cdots},$$

where a_n counts off how many of the numbers $p^0, p^1, \ldots$ are $\leqq n$. Thus $a_n = \left[\frac{\log n}{\log p}\right] + 1$ and consequently $\frac{a_n}{1+\frac{1}{2}+\cdots+\frac{1}{n}} \to \frac{1}{\log p}$, which already completes the proof, according to Probs. 1 and 2. For $b_n = 1 + \frac{1}{2} + \cdots + \frac{1}{n} > 0$, Σb_n diverges, and (see I, § 1, Prob. 13), if we set $1 - z = \rho(\cos\varphi + i\sin\varphi)$,

$$\frac{\Sigma b_n |z|^n}{|\Sigma b_n z^n|} = \frac{|1-z|}{1-|z|} \cdot \frac{\log(1-|z|)}{|\log(1-z)|} \leqq K \cdot \frac{\log\frac{K}{\rho}}{\log\frac{1}{\rho}}.$$

The quotient thus remains less than a fixed bound for all $z \neq 1$ in the triangle $z_1 z_2 1$ which lie sufficiently close to $+1$.

d) Proof according to Probs. 1 and 2. Let $a_0 = 0$, $n^p = a_n$ for $n = 1, 2, \ldots$, and $(1-z)^{-p-1} = \sum_{n=1}^{\infty} b_n z^n$, so that $b_n = \binom{n+p}{n}$. Then $b_n > 0$, Σb_n diverges, and

$$\frac{a_n}{b_n} = \frac{n!n^{p+1}}{(p+1)\cdots(p+1+n)} \cdot \frac{p+1+n}{n} \to \Gamma(p+1).$$

Finally, $\left(\sum_{n=0}^{\infty} b_n |z|^n\right) \Big/ \left|\sum_{n=0}^{\infty} b_n z^n\right| = \left(\frac{|1-z|}{1-|z|}\right)^{p+1} \leqq K^{p+1}$.

6. We have

$$\sum_{n=0}^{\infty} a_n z^n = (1-z)\sum_{n=0}^{\infty} s_n z^n = s + (1-z)\sum_{n=0}^{\infty}(s_n - s)z^n.$$

Hence, if we set $\sqrt{n}(s_n - s) = \varepsilon_n$ with $\varepsilon_n \to 0$ $(n \geqq 1)$, then it is sufficient to show that

$$(*) \qquad \left|(1-z)\sum_{n=1}^{\infty}\frac{\varepsilon_n}{\sqrt{n}}z^n\right| \leqq \frac{|1-z|}{\sqrt{1-|z|}}\cdot\frac{\sum_{n=1}^{\infty}\frac{|\varepsilon_n|}{\sqrt{n}}|z|^n}{\sum_{n=1}^{\infty} b_n|z|^n} \to 0.$$

Here we have set

$$\frac{1}{\sqrt{1-|z|}} = \sum_{n=1}^{\infty} b_n|z|^n, \text{ and hence } b_n = (-1)^n\binom{-\frac{1}{2}}{n}.$$

According to K II, § 6, Ex. 3, (6), footnote 1, $b_n > 0$ and $b_n \cong \dfrac{1}{\sqrt{\pi n}}$ so that $\sum b_n$ diverges. Thus, by Prob. 1, the second factor in (*) tends to 0. It therefore remains to be shown that $\dfrac{|1-z|}{\sqrt{1-|z|}}$ remains bounded for all $z \neq 1$ in the ellipse mentioned in the problem. For these $z = x + iy$, however,

$$\frac{|1-z|^2}{1-|z|} = \frac{(1-x)^2+y^2}{1-\sqrt{x^2+y^2}} \leqq \frac{(1-x)^2+\alpha^2(1-x^2)}{1-\sqrt{x^2+\alpha^2(1-x^2)}}.$$

This $\to \dfrac{2\alpha^2}{1-\alpha^2}$ as $x \to +1$. Therefore the expression remains bounded for the aforementioned z.

Instead of the ellipse referred to in the problem, one may of course also take a circle whose radius is < 1 and which is internally tangent to the unit circle at $+1$.

For a deepening of this problem cf. F. Lösch, *Mathematische Zeitschrift* 37 (1933), pp. 85–89, and W. Meyer-König, *Mathematische Zeitschrift* 46 (1940), pp. 571–590.

7. a) $f_1(z)$ is an ordinary power series, with radius 1, which converges at $z = +1$. Hence, according to Abel's limit theorem (I, § 11, Prob. 10),

$$\lim f_1(z) = 1 + \tfrac{1}{3} - \tfrac{1}{2} + \tfrac{1}{5} + \tfrac{1}{7} - \tfrac{1}{4} + + - \cdots = s.$$

(As is well known, $s = \frac{3}{2}\log 2$; proof?). The limit of the function as $z \to +1$ thus coincides with the value, for $z = +1$, of the series representing the function.

b) For $|z| < 1$, the series for $f_2(z)$ is absolutely convergent, and may therefore be rearranged:

$$f_2(z) = \sum_{n=1}^{\infty} (-1)^{n-1}\frac{z^n}{n} = \log(1+z).$$

Hence, $\lim f_2(z) = \log 2$. The value of the series for $z = +1$, however, is again equal to $s = \frac{3}{2}\log 2$, and is thus different from the limit of the function as $z \to +1$.

8. The given function is equal to

$$\frac{1}{2i}[(1-z)^{-i+1} - (1-z)^{i+1}].$$

The coefficient of z^n in the expansion is equal to the imaginary part of $(-1)^n\binom{-i+1}{n}$, so that its absolute value is less than the absolute value of this binomial coefficient. For $n > 2$, however, this absolute value is equal to

$$\frac{\sqrt{2}}{n(n-1)}\sqrt{\left(1+\frac{1}{1^2}\right)\left(1+\frac{1}{2^2}\right)\cdots\left(1+\frac{1}{(n-2)^2}\right)}$$

(cf. § 5, Prob. 5). This yields the assertion, because the square root tends to a finite limit as $n \to \infty$ (see K II, § 2, Theorem 4).

§ 16. *Analytic Continuation of Power Series*

1. Since $\mathfrak{P}'(z) = \sum_{n=1}^{\infty} z^{n-1} = \frac{1}{1-z}$, $\mathfrak{P}^{(k)}(z) = \frac{(k-1)!}{(1-z)^k}$. Therefore the expansion about the point z_1 reads:

$$\mathfrak{P}(z_1) + \sum_{n=1}^{\infty} \frac{1}{n}\left(\frac{z-z_1}{1-z_1}\right) = \mathfrak{P}(z_1) + \mathfrak{P}\left(\frac{z-z_1}{1-z_1}\right).$$

The radius is again equal to 1, because $|1 - z_1| = 1$, so that, in particular, z_2 falls inside the new circle.

Since $\frac{d}{dz}\mathfrak{P}\left(\frac{z-z_1}{1-z_1}\right) = \frac{1}{1-z_1}\cdot\sum_{n=1}^{\infty}\left(\frac{z-z_1}{1-z_1}\right)^{n-1} = \frac{1}{1-z}$, $\frac{d^k}{dz^k}\mathfrak{P}\left(\frac{z-z_1}{1-z_1}\right) = \frac{(k-1)!}{(1-z)^k}$, and the expansion about the point z_2 reads:

$$\mathfrak{P}(z_1) + \mathfrak{P}\left(\frac{z_2-z_1}{1-z_1}\right) + \mathfrak{P}\left(\frac{z-z_2}{1-z_2}\right).$$

After p steps one obtains the following expansion about the point $z_p = z_0 = 0$:

$$\mathfrak{P}(z_1) + \mathfrak{P}\left(\frac{z_2-z_1}{1-z_1}\right) + \cdots + \mathfrak{P}\left(\frac{z_p-z_{p-1}}{1-z_{p-1}}\right) + \mathfrak{P}\left(\frac{z-z_p}{1-z_p}\right).$$

Since the newly occurring power series is identical with $\mathfrak{P}(z)$, we have shown that $\mathfrak{P}(z)$ can be continued around 1, and, on returning, merely the additive constant

$$\mathfrak{P}(z_1) + \mathfrak{P}\left(\frac{z_2-z_1}{1-z_1}\right) + \cdots + \mathfrak{P}\left(\frac{z_p-z_{p-1}}{1-z_{p-1}}\right)$$

has appeared. Now it is easy to verify that all p arguments here are equal, and hence all are equal to z_1. The constant is therefore equal to

$$p\,\mathfrak{P}(z_1) = p\cdot\left(1 - e^{-\frac{2\pi i}{p}}\right)\left[1 + \frac{z_1}{2} + \frac{z_1^2}{3} + \cdots\right].$$

This value does not depend on p, so that it can be determined by letting $p \to \infty$. Then $z_1 \to 0$, and consequently the bracket $\to 1$, whereas the product preceding it obviously $\to 2\pi i$.

2. Let c be the residue of the pole. Then $f(z) - \frac{c}{z - z_0}$ is regular at z_0 too, and therefore in a circle $|z| < R$ whose radius $R > |z_0| = r$. If we set $f(z) - \frac{c}{z-z_0} = \sum_{n=0}^{\infty} b_n z^n$, then, in particular, $b_n z_0^n \to 0$. Now

$$\frac{c}{z - z_0} = \sum_{n=0}^{\infty} \left(-\frac{c}{z_0^{\,n+1}}\right) z^n, \text{ and hence } a_n = b_n - \frac{c}{z_0^{\,n+1}}.$$

We see now that

$$\frac{a_n}{a_{n+1}} = \frac{b_n z_0^{\,n+1} - c}{b_{n+1} z_0^{\,n+2} - c} \cdot z_0 \to z_0.$$

3. Since $\left(\frac{\zeta}{1-\zeta}\right)^{n+1} = \sum_{k=0}^{\infty} \binom{n+k}{n} \zeta^{k+n+1}$, we have $\mathfrak{P}_1(\zeta) = \sum_{n=0}^{\infty} b_n \zeta^{n+1}$, with $b_n = \binom{n}{0} a_0 + \binom{n}{1} a_1 + \cdots + \binom{n}{n} a_n$. The radius ρ of this power series is given by

$$\frac{1}{\rho} = \limsup \sqrt[n]{\left|\binom{n}{0} a_0 + \binom{n}{1} a_1 + \cdots + \binom{n}{n} a_n\right|}.$$

Consider now the family of circles $|\zeta| = \left|\frac{z}{z+1}\right| = \alpha \geqq 0$. For $\alpha \leqq \frac{1}{2}$, these circles lie wholly within the unit circle, but project beyond the latter for $\alpha > \frac{1}{2}$. Let $f(z)$ be the function defined by the series $\mathfrak{P}(z)$ and its rectilinear continuation (from the origin). Then the largest of these circles (i.e., that circle which corresponds to the largest α-value; its interior is to be regarded as that part of the plane which contains the origin, so that for $\alpha > 1$ the interior also contains the point ∞) which is still free of singular points of the function $f(z)$ yields, through its α-value, the radius ρ of $\mathfrak{P}_1(\zeta)$. Hence, $\frac{1}{2} \leqq \rho \leqq +\infty$.

4. Obviously the point $+1$ is singular if, and only if, the circle $\alpha = \frac{1}{2}$ is the largest of the circles mentioned in the preceding problem, and hence, if, and only if,

$$\limsup \sqrt[n]{\left|\binom{n}{0} a_0 + \cdots + \binom{n}{n} a_n\right|} = 2.$$

And the point is therefore regular if, and only if, this lim sup < 2 (since it certainly cannot be > 2).

5. a) Here we are concerned with

$$\limsup \sqrt[n]{\left|\binom{n}{0} - \binom{n}{1} + \cdots + (-1)^n \binom{n}{n}\right|}\,.$$

Since the radicand $= (1 - 1)^n = 0$, the lim sup $= 0 < 2$, and hence the point $+1$ is regular. (In fact, the function represented is equal to $\dfrac{z}{1+z} = \zeta$, and is thus an *entire* function of ζ.)

b) Here we are interested in

$$\limsup \sqrt[n]{\left|\binom{n}{0} - \tfrac{1}{2}\binom{n}{1} + \tfrac{1}{3}\binom{n}{2} - \cdots + \frac{(-1)^n}{n+1}\binom{n}{n}\right|}\,.$$

The radicand is the nth difference of the sequence $1, \frac{1}{2}, \frac{1}{3}, \ldots$, and hence (as is easily verified) $= \dfrac{1}{n+1}\cdot$ Therefore the lim sup $= 1$. Consequently, the function represented by the series is regular in the circle $\Re(z) > -\frac{1}{2}$ of the family, and, in particular, at $+1$. The "larger" circles contain the point ∞, so that the function is no longer regular in them.

6. We may assume that *all* $a_n \geqq 0$ and that the radius of $\sum a_n z^n$ is equal to 1. Then, for a given $\varepsilon > 0$, we have $a_n > (1 - \varepsilon)^n$ infinitely often. If $n = m$ is such an index, then

$$\sqrt[2m]{\binom{2m}{0} a_0 + \cdots + \binom{2m}{m} a_m + \cdots + \binom{2m}{2m} a_{2m}} \geqq \sqrt[2m]{\binom{2m}{m}}\,(1 - \varepsilon)^m.$$

Since $\dbinom{2m}{m} \cong \dfrac{2^{2m}}{\sqrt{\pi m}}$ (cf. K II, p. 50, footnote 1), the lim sup of our radical is $\geqq 2$, and hence is $= 2$.

7. According to the theorem proved in I, § 11, Prob. 3, $+1$ is a singular point for the function $\varphi(z)$. Therefore the Taylor expansion of $\varphi(z)$ about the point $+\frac{1}{2}$, that is, the series

$$\sum_{\nu=0}^{\infty} \frac{\varphi^{(\nu)}(\frac{1}{2})}{\nu!}(z - \tfrac{1}{2})^\nu,$$

diverges for all real $z > 1$. Since $\varphi^{(\nu)}(\frac{1}{2}) = \Re(f^{(\nu)}(\frac{1}{2}))$, the series

$$\sum_{\nu=0}^{\infty} \frac{f^{(\nu)}\left(\frac{1}{2}\right)}{\nu!} \left(z - \tfrac{1}{2}\right)^{\nu}$$

is also divergent for all real $z > 1$. Consequently, $f(z)$ is singular at $+1$.

§ 17. *Analytic Continuation of Arbitrarily Given Functions*

1. Obviously not, for it is not differentiable at $x = 0$. Or: For $x > 0$, $F(x) = x$ and is therefore continuable, and $F(z) = z$ is the continuation. Analogously, $F(x)$ is also continuable for $x < 0$, and the continuation is $F(z) = -z$. A function, however, cannot be continued in two distinct ways (see K I, § 22, Theorem 2).

2. Not in the sense of K I, p. 95 that there exists a region $\mathfrak{G}$ containing the segment $-1 < x < +1$ in its interior, and a function $f(z)$, regular in $\mathfrak{G}$, whose values along the segment mentioned coincide with the values given there. For, as the course of values along $0 < x < +1$ shows, only e^{-1/z^2} comes into consideration as a possible continuation, but this function has an essential singularity at 0.

3. We see first, according to I, § 9, Prob. 1f or from $\left|\dfrac{z^n}{1 - z^n}\right| \leqq \dfrac{1}{1 - \rho} \cdot \rho^n$ for $|z| \leqq \rho$, that $f(z)$ is regular in $|z| < 1$. Let ζ be a primitive root of unity of degree $g_0 \cdot g_1 \cdots g_p$ ($p \geqq 0$ fixed). Then, for $0 < \rho < 1$, $z = \rho\zeta$,

$$\sum_{n=p}^{\infty} \frac{z^{g_0 \cdots g_n}}{1 - z^{g_0 \cdots g_n}} = \sum_{n=p}^{\infty} \frac{\rho^{g_0 \cdots g_n}}{1 - \rho^{g_0 \cdots g_n}}.$$

This increases (remaining positive) beyond all bounds as $\rho \to 1$. The omitted initial part of the series is a rational function which is regular at ζ, and therefore remains bounded as $z \to \zeta$. Consequently, ζ is a singular point of $f(z)$. Since the points ζ lie everywhere dense on $|z| = 1$, $f(z)$ is not continuable beyond the unit circle.

4. Let $f(z)$ be regular for $|z| < r$, $r > 0$. Then (*) shows that $f(2z)$ is also regular for $|z| < r$, and hence $f(z)$ is regular for $|z| < 2r$; consequently, also for $|z| < 4r$, $< 8r$, ..., and therefore in the entire

plane. Obviously the only essential property of the right-hand side of (*) is that it is an entire rational function of $f(z)$ and its derivatives.

5. For $\Re(z) > 0$ it follows from (†) that

$$(\dagger\dagger)\quad f(z+k+1) = (z+k)(z+k-1)\cdots(z+1)z\cdot f(z),$$
$$k \geqq 0 \text{ integral.}$$

Now $f(z+k+1)$ is regular for $\Re(z) > -k-1$, and therefore so is the right-hand side. Hence, for $\Re(z) > -k-1$, $f(z)$ has no singularities other than simple poles at $0, -1, -2, \ldots, -k$. For a neighborhood of $-k$, $f(z)$ has the form $\dfrac{r_k}{z+k} + f_k(z)$, where r_k denotes the residue at $z = -k$ and $f_k(z)$ is a function which is regular at $-k$. Now letting $z \to -k$, (††) yields

$$f(1) = (-1)^k k!\, r_k, \qquad r_k = \frac{(-1)^k}{k!} f(1).$$

6. Let z_0 be a point of $\mathfrak{w}$, and let $\mathfrak{C}$ be a circle about z_0 which lies in $\mathfrak{G}$ and intersects $\mathfrak{w}$ exactly twice. The region enclosed by $\mathfrak{C}$ is then cut by $\mathfrak{w}$ into two parts, $\mathfrak{G}_1$ and $\mathfrak{G}_2$, whose boundaries we shall call $\mathfrak{C}_1$ and $\mathfrak{C}_2$. Now it is not difficult to verify that Cauchy's integral formula (K I, § 15) is also valid in this case in which we know about $f(z)$ merely that it is regular inside the boundary curve (consisting of a circular arc and a finite number of segments) and assumes continuous boundary values along this curve. Consequently, for z in $\mathfrak{G}_1$,

$$\frac{1}{2\pi i}\int_{\mathfrak{C}_1} \frac{f(\zeta)}{\zeta - z}\, d\zeta = f(z), \quad \frac{1}{2\pi i}\int_{\mathfrak{C}_2} \frac{f(\zeta)}{\zeta - z}\, d\zeta = 0.$$

Hence, $f(z) = \dfrac{1}{2\pi i}\displaystyle\int_{\mathfrak{C}} \frac{f(\zeta)}{\zeta - z}\, d\zeta$. It can be shown in an analogous fashion that this last formula is also valid for every z in $\mathfrak{G}_2$. According to K I, § 16, however, the integral defines a function which is regular at every point of the region enclosed by $\mathfrak{C}$. Therefore $f(z)$ is also regular at z_0. (Is the assertion true for more general paths $\mathfrak{w}$?)

7. According to I, § 5, Prob. 9, where of course instead of the unit

circle a broken line may be taken as the boundary curve, the boundary values form a continuous function along $\mathfrak{w}$. The theorem of the preceding problem now immediately completes the proof.

8. No, because the region under consideration is doubly connected, not simply connected. Cf. § 18, Prob. 7.

9. The algebraic function defined by means of $w^3 - 3w = z$ (cf. § 19, Prob. 4d) is triple-valued and has branch-points at $+2$, -2, and ∞. At $+2$, only the first and third sheets hang together, the second is smooth; at -2, however, the first and second sheets hang together, whereas the third is smooth. Let us take $\mathfrak{G}$ to be the circle $|z| < 4$, and for $f(z)$ choose one of the three branches of this function which are regular in a neighborhood of $z_0 = 0$. Then $f(z)$ can be continued to $+2$ and to -2. For if the given branch $f(z)$ lies on the first sheet, an encirclement of -2 leads to the second sheet, on which one can now proceed unhindered to $+2$. If $f(z)$ lies on the third sheet, then successive encirclements of $+2$ and -2 again lead to the second sheet, on which one can again proceed to $+2$. If $f(z)$ lies on the second sheet, one can proceed immediately to $+2$. An analogous treatment holds for -2. One can reach every other point of $\mathfrak{G}$ along any path which avoids ± 2.

10. The proof runs exactly the same as that of the monodromy theorem itself (see K I, § 25), except that now a point is not to be considered "singular", i.e., a hindrance to the continuation process, unless neither the function nor its reciprocal can be continued to the point in question.

11. The proof again runs the same, except that, if a point ζ hinders the continuation process, the domain of values obtained in $0 < |z - \zeta| < \delta$ must be represented by a Laurent expansion.

CHAPTER VI

MULTIPLE-VALUED FUNCTIONS AND RIEMANN SURFACES

§ 18. *Multiple-valued Functions in General*

1. Obviously yes; e.g., log z in $1<|z|<2$. Not in a simply-connected region, however, which is indeed precisely what the monodromy theorem asserts (K I, § 25).

2. a) $(z-1)\sqrt{z}$ has the value 0 at the point $+1$ of each of the two sheets. b) $(e^z-1)\sqrt{z}$ has the value 0 at the points $z=2k\pi i$, $k=0, \pm 1, \pm 2, \ldots$, of both sheets. c) At all points of two superposed regions, naturally not, because then the functional branches spread out on the two sheets would be completely identical, so that these sheets could not have been constructed as two distinct sheets in the first place.

3. a) Two single-valued functions, namely, $e^{\frac{z}{2}}$ and $-e^{\frac{z}{2}}$, whose domains of values must be spread out on two completely separate sheets.

b) A two-valued function, whose domain of values must be spread out on a two-sheeted surface which is branched at the points $(2k+1)\frac{\pi}{2}$ $(k=0, \pm 1, \pm 2, \ldots)$. An encirclement of one of these points invariably leads from one sheet to the other.

c) Two single-valued functions, namely, $+\cos z$ and $-\cos z$. (Reason: All zeros of the radicand are zeros of order two.)

d) Two single-valued functions, because all poles and all zeros of the radicand are of the second order (the latter because $\wp'(\frac{1}{2}\omega)=0$). The radical thus yields two single-valued functions in a neighborhood of every point of the plane. Hence, according to § 17, Prob. 10, we have two single-valued functions for the entire plane.

e) If the two zeros of $\wp(z)$ — concerning whose position one cannot assert much in general — are distinct, then we obtain a two-valued function. If, however, $\wp(z)$ has only *one* zero of order two in the parallelogram (which can only occur if $\wp(z)$ vanishes for $\frac{1}{2}\omega$ or $\frac{1}{2}\omega'$ or $\frac{1}{2}(\omega + \omega')$), then we obtain two single-valued functions, as in d).

f) Infinitely many single-valued functions, namely, $z + 2k\pi i$, $k = 0, \pm 1, \pm 2, \ldots$.

g) An infinitely multiple-valued function with branch-points at $0, \pm\pi, \pm 2\pi, \ldots$.

h) A single-valued function, as the power-series expansion shows.

4. Choose a definite path $\mathfrak{k}_0$ from 0 to z_0. Let the value of the integral obtained be J_0.

Every other path $\mathfrak{k}$ running from 0 to z_0 in the simple z-plane can be written in the form $(\mathfrak{k} - \mathfrak{k}_0) + \mathfrak{k}_0$, i.e., as a closed path from 0 to 0 and from here along $\mathfrak{k}_0$ to z_0. We are therefore concerned merely with evaluating the integral along closed paths. Now such a path (provided that it avoids ± 1) may be replaced by a succession of "loops" (i.e., paths which lead from 0 to a neighborhood of $+1$ (or -1), encircle this point once, and return to 0) without altering the value of the integral. (This can be shown with complete rigor by means of considerations analogous to those in § 1, Prob. 6.) If we observe that the radical changes its sign on encircling ± 1, that the "loop integral" about $+1$ has the value $2\int_0^1 \frac{dx}{\sqrt{1-x^2}}$ whereas that about -1 has the value $-\pi$, irrespective of whether the loop is described positively or negatively (!), then we see that two loops described in succession yield either 2π or 0 or -2π. Therefore if we have an even number of loops, the integral over the whole closed curve is equal to $2k\pi$, $k = 0, \pm 1, \pm 2, \ldots$; if the number is odd, the integral equals $2k\pi + \pi$. In the first case we return to 0 with the original value of the radical, in the second, with its negative. Hence, the value of the integral taken along $\mathfrak{k}$ is equal to

$$2k\pi + J_0, \quad 2k\pi + \pi - J_0,$$

respectively. For $z = \pm 1$, if $\mathfrak{k}_0$ is taken to be a rectilinear segment, $J_0 = \pm\frac{\pi}{2}$; for $z_0 = \infty$ the integral diverges.

5. a) $\int_1^{z_0} \frac{dz}{\sqrt[p]{z}} = \frac{p}{p-1}\left[\frac{z}{\sqrt[p]{z}}\right]_1^{z_0}$, provided that the values of $\sqrt[p]{z_0}$, $\sqrt[p]{1}$ corresponding to the respective sheets are taken here.

b) $\int_1^{z_0} \log z\, dz = [z(\log z - 1)]_1^{z_0}$, if $\log z_0$ and $\log 1$ are taken in an analogous manner (see a)).

6. a) According to I, § 11, Prob. 4d, $f(z)$ is regular in $|z| < 1$ but not continuable across the boundary $|z| = 1$. According to § 20, Prob. 13, $f(z)$ assumes in $|z| < 1$ no value more than once. Since $f(0) = 0$, $f(z) \neq 0$ for all other z. Therefore $\frac{f(z)}{z}$ is regular and $\neq 0$ in $|z| < 1$, and consequently two single-valued, regular functions are represented by $\sqrt{\frac{f(z)}{z}}$ in $|z| < 1$. $F_1(z) = \sqrt{z} \cdot \sqrt{\frac{f(z)}{z}}$ is thus a two-valued function in $|z| < 1$, whose Riemann surface is that part of the surface for $\sqrt{z}$ whose points lie in the interior of the unit circle.

b) One finds in an entirely analogous manner that the Riemann surface of $F_2(z)$ is that part of the surface for $\log z$ whose points lie in the interior of the unit circle.

7. The function $\log f(\frac{1}{2}z)$ is regular in $0 < |z| < 2$ and behaves like $\log z$ with respect to multiple-valuedness; $\log f\left(\frac{1}{2z}\right)$ is regular in $0 < \left|\frac{1}{2z}\right| < 1$, i.e., in $\frac{1}{2} < |z| < +\infty$, and behaves like $\log \frac{1}{z}$ with respect to multiple-valuedness. Therefore $G_1(z)$, the sum of the two, is single-valued and regular in $\frac{1}{2} < |z| < 2$. (More precisely: The expression for $G_1(z)$ defines infinitely many single-valued, regular functions, which differ, however, merely by additive constants of the form $2k\pi i$.) It is, moreover, not continuable across the boundaries of the ring. The function $G_2(z)$ is also regular in the same ring, is not continuable across its boundaries, but alters its value by $4\pi i$ when

continued inside the ring once around the inner disk $|z| \leq \frac{1}{2}$ in the positive sense. (Cf. Prob. 1 and § 17, Prob. 8.) The Riemann surface for $G_2(z)$ is thus that part of the surface for log z whose points satisfy the relation $\frac{1}{2} < |z| < 2$.

§ 19. *Multiple-valued Functions; in Particular, Algebraic Functions*

Preliminary remarks to Probs. 1 *and* 2. The problems can be handled quite simply and uniformly. For every function one can tell immediately how-many-valued it is, and consequently how many sheets the Riemann surface has, what the difference is between the various values of the function for the same z (so that all the values can be written down immediately with the aid of *one* of them), and which points a, b, ... come into consideration as the only possible branch-points. Now take as many copies of the z-plane as the multiple-valuedness indicates, mark on them the points which come into consideration as branch-points (counting ∞ among them too), and join them in any order, the last with ∞, by means of straight line segments, say, in such a manner that they do not intersect one another, which can always be brought about by altering the order of succession or the connecting line. The copies are cut along these connecting lines. On a first copy choose a point z_0 which does not lie on the cut, expand a branch of the function (which is then regular at z_0) for a neighborhood of z_0, and continue this functional element in the plane without crossing the cut. In this way, every point of the sheet is assigned a functional value; along the edges of the cut we still obtain continuous boundary values, but, in general, different ones along the two banks. The covering of the remaining sheets, according to the remark made above, is given automatically. We see likewise how the coverings of the two banks of the same cut differ. Now pile up the sheets in a definite order and fuse the identically covered cut-edges.

In detail this is done as follows: The multiple-valuedness comes about in the following problems solely through the behavior of am $(z-a)$, am $(z-b)$, ... This behavior, however, can be surveyed immediately by letting z describe curves which remain in the vicinity of the cut-edges, encircle certain of the points a, b, ..., and leave the others outside. Then it becomes immediately clear how the functional values differ at the two cut-edges; for in general am $(z - z_1)$ increases by $+2\pi$ if z_1 is encircled in the positive sense. The cut-edges of the various sheets are to be fused accordingly.

We make the following agreement as to notation for this fusion: If, in the vicinity of a, we cross a certain cut, emanating from a, in such a direction that a lies to the left, and if, in doing so, sheet 1 is to be joined to sheet α_1, sheet 2 to sheet α_2, ..., then we express this by means of the symbol

$$\begin{pmatrix} 1 & 2 & \cdots \\ \alpha_1 & \alpha_2 & \cdots \end{pmatrix}_a$$

assigned to this cut. If we describe in this way the manner of fusion at all cut-edges, the surface is completed.

1. a) Three sheets. There is a three-sheeted branch-point at a and at ∞. Superposed functional values differ merely by the factor ω^k with $k = 0, 1, 2$ and $\omega = e^{\frac{2\pi i}{3}}$. If w is affixed to the first sheet, ωw to the second, and $\omega^2 w$ to the third, then the sheets are joined according to $\binom{1\,2\,3}{2\,3\,1}_a$.

b) Three sheets. There is a three-sheeted branch-point at a and at ∞. Encirclement of a once produces the factor ω^2. Consequently, if we affix w to the first sheet, $\omega^2 w$ to the second, and $\omega^4 w = \omega w$ to the third, then the sheets are again joined according to $\binom{1\,2\,3}{2\,3\,1}_a$. The surface is then identical with the preceding one. One may, however, also affix ωw to the second sheet and $\omega^2 w$ to the third; then the joining takes place according to $\binom{1\,2\,3}{3\,1\,2}_a$.

c) Three sheets. Possible branch-points a, b, ∞. Superposed functional values differ merely by ω^k. If we affix w, ωw, $\omega^2 w$ to the first, second, third sheets, respectively, then the sheets are joined along the cut $a \ldots b$ according to $\binom{1\,2\,3}{2\,3\,1}_a$. For, if a is encircled, am $(z - a)$ increases by $+2\pi$ whereas am $(z - b)$ does not change. If a and b are encircled simultaneously, am $(z - a)$ *and* am $(z - b)$ increase by $+2\pi$, so that their difference remains unchanged. On every single sheet we close the cut, the cut $b \ldots \infty$ therefore disappears, ∞ is not a branch-point, the sheets are piled up separately there. (Visualize this with the aid of the Riemann *sphere*.) We have thus a three-sheeted surface with two three-sheeted branch-points, at a and b.

d) We begin exactly as in c). Along $a \ldots b$ the sheets are joined again according to $\binom{1\,2\,3}{2\,3\,1}_a$, along $b \ldots \infty$, however, according to $\binom{1\,2\,3}{3\,1\,2}_b$. We obtain a three-sheeted surface with three three-sheeted branch-points, at a, b, ∞.

e) Exactly as in d) if we introduce the point c instead of ∞. At ∞ the sheets are not joined, so that ∞ in this case is not a branch-point.

f) A three-sheeted surface. Possible branch-points at a_1, a_2, ..., a_k and ∞. Considerations analogous to those in c) and d) show that along $a_1 \ldots a_2$ the sheets are joined according to $\binom{1\,2\,3}{2\,3\,1}_{a_1}$, along $a_2 \ldots a_3$ according to $\binom{1\,2\,3}{3\,1\,2}_{a_2}$, along $a_3 \ldots a_4$ according to $\binom{1\,2\,3}{1\,2\,3}_{a_3}$. This third cut therefore disappears. Along $a_4 \ldots a_5$ the sheets are again joined according to $\binom{1\,2\,3}{2\,3\,1}_{a_4}$, etc. The point ∞ is a branch-point (and hence the cut $a_k \ldots \infty$ is not left out) if, and only if, k is not divisible by 3.

g) Six sheets. Possible branch-points at a, b, c, ∞. The difference between the six functional values at a regular point z is that ω^k can appear as factor of the cube root, -1 as factor of the square root. Consequently, if w_1 is a value of the cube root and w_2 is a value of the square root, we affix the values $w_1 + w_2$, $\omega w_1 + w_2$, $\omega^2 w_1 + w_2$, $w_1 - w_2$, $\omega w_1 - w_2$, $\omega^2 w_1 - w_2$ in succession to sheets 1 to 6. If we encircle a alone, only am $(z - a)$ changes; the sheets are joined along $a \ldots b$ according to $\binom{1\,2\,3\,4\,5\,6}{2\,3\,1\,5\,6\,4}_a$. (We see that the first three sheets, as well as the last three sheets, hang together in a "cycle of degree 3"; i.e., by encircling a alone we always remain either in the one or the other of the two groups of sheets.) If a and b are encircled simultaneously, the amplitude of the radicand of the cube root decreases by 2π; hence, along $b \ldots c$ the sheets are joined according to $\binom{1\,2\,3\,4\,5\,6}{3\,1\,2\,6\,4\,5}_b$, which again gives two third degree cycles. If all three points are encircled simultaneously, the cube root behaves in the same way but the square root changes sign. Hence, along $c \ldots \infty$ the sheets are joined according to $\binom{1\,2\,3\,4\,5\,6}{6\,4\,5\,3\,1\,2}_c$. We thus have a six-sheeted surface with four branch-points, at a, b, c, ∞. By encircling also b, c and ∞ *individually*, we find more precisely that at each of the points a and b there are two superposed three-sheeted branch-points; the first three sheets form a cycle and the last three sheets form a cycle. At c, three two-sheeted branch-points are superposed; the first and fourth, the second and fifth, and the third and sixth sheets from cycles of degree two. At ∞ all six sheets hang together and form a single cycle of degree six.

h) An n-sheeted surface. The functional values differ by factors which are powers of $\omega = e^{\frac{2\pi i}{n}}$. If we affix w to the first sheet, ωw to the second, ..., $\omega^{n-1} w$ to the nth, then the usual considerations show that the sheets are joined along $a \ldots b$ according to $\begin{pmatrix}1, 2, \ldots, n-1, n \\ 2, 3, \ldots, \quad n, \quad 1\end{pmatrix}_a$, along $b \ldots c$ according to $\begin{pmatrix}1, \ldots, n-2, n-1, n \\ 3, \ldots, \quad n, \quad 1, \quad 2\end{pmatrix}_b$, and along $c \ldots \infty$ according to $\begin{pmatrix}1, 2, \ldots, n \\ 4, 5, \ldots, 3\end{pmatrix}_c$. We thus obtain four n-sheeted branch-points, at a, b, c, ∞. A positive encirclement of a or b or c leads in each case to the next sheet (the nth sheet naturally is "followed" by the 1st), a positive encirclement of ∞ (i.e., one which keeps ∞ to the left) leads from the kth sheet to the $(k-3)$d.

2. a) Infinitely many sheets. Branch-points at a and at ∞. If we spread out a branch of the function, which we denote by w, on a zeroth sheet in accordance with the preliminary remark, then $w + 2k\pi i$, $k = \pm 1, \pm 2, \ldots$, are affixed to the other sheets. We number the sheets with the values of k, and place those with larger k above those with smaller k. The sheets are then joined according to $\begin{pmatrix}k \\ k+1\end{pmatrix}_a$. The surface is obviously constructed like the one for $\log z$ except that the branch-points lie at a and ∞ instead of at 0 and ∞.

b) Infinitely many sheets. Possible branch-points at a, b and ∞. Functional values are again $w + 2k\pi i$, $k = 0, \pm 1, \pm 2, \ldots$, which are affixed analogously to the kth sheet. The sheets are again joined along

$a \ldots b$ according to $\begin{pmatrix} k \\ k+1 \end{pmatrix}_a$, but along $b \ldots \infty$ according to $\begin{pmatrix} k \\ k+2 \end{pmatrix}_b$, because if a and b are encircled simultaneously, both am $(z - a)$ and am $(z - b)$ increase by 2π.

c) Similar to b). Since, however, am $\left(\frac{z-a}{z-b}\right)$ does not change when a and b are encircled simultaneously, the cuts along $b \ldots \infty$ in each individual sheet must be closed. There is no branch-point at ∞, the sheets are separate there.

d) $\log (1 + z^2) = \log (z - i)(z + i)$ can be treated according to b).

e) $\arctan z = \frac{i}{2} \log \frac{1+z}{1-z}$ is handled according to b).

3a) Since $z^i = e^{i \log z}$, precisely the surface for $\log z$. A single positive encirclement of the origin furnishes the functional value with the factor $e^{-2\pi}$.

b) Likewise. If we set $a = \alpha + i\beta$, the factor appearing is $e^{-2\beta\pi + 2\alpha\pi i}$.

4. a) $w = \sqrt[3]{z+1}$. The surface can be constructed according to Prob. 1a.

b) We have $w = \frac{1}{2}(z + \sqrt{(z-2)(z+2)})$. Hence, according to K II, § 12, the surface is two-sheeted with branch-points at $+2$ and -2. A single encirclement of exactly one of these points leads from w to $-w + z = \frac{1}{w}$, and hence a double encirclement leads back to w.

Instead of this direct treatment one can also give the following: The (implicitly) given function is the inverse of $z = z(w) = w + \frac{1}{w}$, which is investigated in detail (with letters interchanged) in B, § 12. If w runs rectilinearly from $-\infty$ to -1, then z runs likewise from $-\infty$ to -2. If w then proceeds from -1 to $+1$ along the upper half of the unit circle, z continues rectilinearly from -2 to $+2$. If, finally, w moves rectilinearly from $+1$ to $+\infty$, z goes likewise from $+2$ to $+\infty$. Consequently (cf. § 20, Probs. 3 and 4, and B, § 9) the region

$|w| > 1$ in the upper half-plane is mapped on the upper half of the z-plane. Similarly (or, more simply, according to the reflection principle, B, § 10) one finds that the region $|w| < 1$ in the upper half-plane is mapped on the lower half of the z-plane. To every point z of the z-plane we now imagine that value of w to be affixed for which this z is obtained. This is one of the two values of the given function $w = w(z)$, and, in fact, that one for which $\mathfrak{J}(w) \geqq 0$. The two rays $-\infty \cdots -2$ and $+2 \cdots +\infty$ are then covered twice. We accordingly cut the z-plane along these rays and cover both banks. (Or, we can cover the two halves of the z-plane separately and then join them along the segment $-2 \cdots +2$ of the boundaries, which alone is covered with the same values on both halves.)

In exactly the same way (or again by reflection in the real axis of the w-plane) we find that the lower half of the w-plane is also mapped on such a z-sheet. In this case the *inner* semicircle of the unit circle in the w-plane is mapped on the *upper* half of the z-plane, and the outer semicircle is mapped on the lower half. In this z-sheet too we attach to every point z that value w for which this z is obtained.

We place one z-sheet over the other, and join them by fusing those banks of the cuts which are covered with the same functional values; stated simply, we join them "crosswise". This completes the surface. If the value w is affixed to the point z of one of the sheets, then the value affixed to the same point z of the other sheet is obtained by first reflecting w in the unit circle and then reflecting the resulting point in the real axis. The final value is thus $\dfrac{1}{w}$.

c) We proceed as in the preceding problem and first investigate the inverse function $z = z(w) = w^n + w^{-n}$. After the discussion of the preceding problem, the following is immediately clear (make detailed sketches again): If we consider the sector $0 \leqq \text{am } w \leqq \dfrac{\pi}{n}$, and if w describes the boundary of that part of the sector which lies outside $|w| = 1$, beginning at ∞ and proceeding so that the interior of this region remains to the left, then z describes the real axis from left to right. If w likewise describes the boundary of that part of the sector lying inside $|w| = 1$, beginning at 0, then z describes the real axis from right to left. Hence — the argument is exactly the same as in b) — the entire sector is mapped in a one-to-one manner on precisely such a z-sheet as was considered in connection with b). The sectors $(k-1)\frac{\pi}{n} \leqq \text{am } w \leqq k\frac{\pi}{n}$, $k = 1, 2, \ldots, 2n$, are mapped in an entirely similar manner on z-sheets of the same kind, which we number according to the value of k. At every point z of every sheet we now imagine once more that value w to be affixed for which this z is obtained. We pile up these $2n$ sheets (the first on the bottom, say) and join them by fusing the banks which bear the same functional values. The figure in the w-plane, in which the rays $\text{am } w = (k-1)\frac{\pi}{n}$, $k = 1, 2, \ldots, 2n$, alternately yield the rays $+2 \ldots +\infty$ and $-2 \ldots -\infty$ (each twice), now shows immediately that the first z-sheet is to be joined along $-\infty \ldots -2$ to the second sheet, along $+2 \ldots +\infty$ to the $(2n)$th sheet. In general, every sheet with an odd number is joined along $-\infty \ldots -2$ to the succeeding sheet, along $+2 \ldots +\infty$ to the preceding sheet. With even-numbered sheets we do the reverse. Thus, according to the notation given in the preliminary remarks to Probs. 1 and 2, the $2n$ sheets are joined along the cuts $-2 \ldots -\infty$, $+2 \ldots +\infty$ according to

$$\begin{pmatrix} 1, 2; 3, 4; \ldots; 2n-1, & 2n \\ 2, 1; 4, 3; \ldots; \quad 2n, & 2n-1 \end{pmatrix}_{(-2)},$$

$$\begin{pmatrix} 1; 2, 3; \ldots; 2n-2, 2n-1; 2n \\ 2n; 3, 2; \ldots; 2n-1, 2n-2; 1 \end{pmatrix}_{(+2)},$$

respectively. A positive encirclement of both points simultaneously, or what is the same, a negative encirclement of ∞, leads from sheet 1 to sheet 3, from 3 to 5, ..., from $2n-1$ to 1, and likewise from 2 to 4, ..., from $2n-2$ to $2n$, from $2n$ back to 2. Whereas n two-sheeted branch-points (n cycles of degree two) are situated above $+2$ and above -2, there are two n-sheeted branch-points (two cycles of degree n) at ∞. (That these cycles penetrate one another in the described manner naturally makes no difference whatsoever. Cf. K II, p. 103, footnote.)

d) As in the two preceding problems, we first consider the inverse function $z = z(w) = w^3 - 3w$. In order to find out which regions in the w-plane are mapped on the upper and lower halves of the z-plane, we set $w = u + iv$, which yields

$$\Re(z) = u(u^2 - 3v^2 - 3) \text{ and } \Im(z) = v(3u^2 - v^2 - 3).$$

Thus, z is real for $v = 0$ and for $u^2 - \frac{1}{3}v^2 = 1$. Now the real axis and this hyperbola divide the w-plane into six regions, which we number as follows: Between the branches of the hyperbola, below 1, above 1′; inside the right branch of the hyperbola, above 2, below 2′; inside the left branch, above 3, below 3′. (Make a detailed sketch again.) If we now let w traverse the boundaries of these regions, always beginning at ∞ and proceeding so that the encircled region lies on the left in each case, we find that the unprimed regious yield the upper half of the z-plane, the primed regions yield the lower half. We accordingly take three z-sheets, cut them along the real axis, and call the upper half-planes 1, 2, 3, the lower ones 1′, 2′, 3′. To every point z of each of these half-planes we affix that value w whose image it is under the described mapping. (For every z, this is a certain one of the three roots of $w^3 - 3w - z = 0$.) As the figure in the w-plane immediately shows, sheets 1 and 1′ then bear the same values along the boundary segment $-2 \ldots +2$ and are joined accordingly, so that $-\infty \ldots -2$ and $+2 \ldots +\infty$ remain cuts. Sheets 2 and 2′ bear the same values along that boundary segment which is obtained for $1 \leqq w \leqq +\infty$, namely, $-2 \ldots 0 \ldots +\infty$, and are joined there accordingly; $-\infty \ldots -2$ remains a cut. Similarly, 3 and 3′ are joined along $-\infty \ldots 0 \ldots +2$; $+2 \ldots +\infty$ remains a cut. The figure in the w-plane now shows further how these three z-sheets are to be joined. In the customary notation, the three sheets hang together along $-\infty \ldots -2$, $+2 \ldots +\infty$, according to

$$\begin{pmatrix} 1\,2\,3 \\ 2\,1\,3 \end{pmatrix}_{(-2)}, \quad \begin{pmatrix} 1\,2\,3 \\ 3\,2\,1 \end{pmatrix}_{(+2)},$$

respectively. Thus, in each case, one sheet passes smoothly over the point. If we encircle both points simultaneously in the positive sense (or what is the same, ∞ alone in the negative sense), the first sheet goes over into the second, the second into the third, the third into the first; thus all three sheets hang together at ∞ in a single cycle.

e) If we replace w by iw and at the same time z by $-iz$, then the presented connection between z and w goes over into $w^3 - 3w - z = 0$, i.e., into the connection discussed in d). In relation to d) one has therefore merely to carry out a rotation through $\pm\frac{1}{2}\pi$ in the z-plane and in the w-plane in order to survey the present position.

CHAPTER VII

CONFORMAL MAPPING

§ 20. *Concept and General Theory*

1. a) Through am $\dfrac{w_1 - w_0}{z_1 - z_0}$; i.e., the vector extending from z_0 to z_1 must be rotated through this angle in the positive sense in order to become parallel to, and have the same direction as, the vector extending from w_0 to w_1.

b) In the ratio $\left|\dfrac{w_1 - w_0}{z_1 - z_0}\right|$.

2. a) According to the general agreements in K I, § 32 and B, § 3, if, and only if, $1/f(z)$ maps a neighborhood of z_0 conformally on a neighborhood of the origin. If $f(z)$ has a pole of order β at z_0, $1/f(z)$ has a zero of order β there. Hence, there is conformality if, and only if, the pole of $f(z)$ at z_0 is simple. b) Here we have to examine $1/f\left(\dfrac{1}{z'}\right)$ at $z' = 0$. If, however, $f(z)$ has a pole of order β at ∞, then this new function has a zero of order β at $z' = 0$. Hence there is conformality if, and only if, the pole is simple. c) There is conformality if, and only if, $z = \infty$ is a w_0-point of order unity.

3. The theorem and its proof are very similar to those presented in B, § 9. Let $\mathfrak{G}'$ be the set of values $w = f(z)$ for z in $\mathfrak{G}$ and on $\mathfrak{C}$. Let ω be a boundary point of this set. Then there exists a sequence of points $\{w_n\}$ in $\mathfrak{G}'$ tending to ω. Let the z_n be chosen so that $f(z_n) = w_n$. They lie in $\mathfrak{G}$ or on $\mathfrak{C}$ and therefore have at least one limit point ζ there. We must have $f(\zeta) = \omega$. Therefore ζ cannot lie inside $\mathfrak{G}$, because otherwise a full neighborhood of ζ would be mapped on a full neighborhood of ω, whereas ω was supposed to be a boundary point. Consequently, ζ lies on $\mathfrak{C}$, and hence ω lies on $\mathfrak{C}'$. But then $\mathfrak{G}'$ has

no point whatsoever outside $\mathfrak{C}'$. For suppose that w_0 of $\mathfrak{G}'$ were exterior to $\mathfrak{C}'$. Let $\mathfrak{p}$ be a polygonal line connecting w_0 with ∞ without meeting $\mathfrak{C}'$. Since $f(z)$ is bounded in $\mathfrak{G}$, there must be a boundary point of $\mathfrak{G}'$ on $\mathfrak{p}$, which is impossible, since $\mathfrak{p}$ does not intersect the curve $\mathfrak{C}'$. Therefore $\mathfrak{G}'$ lies entirely inside and on $\mathfrak{C}'$. If w_1 now is an arbitrary point inside $\mathfrak{C}'$, we have

$$1 = \frac{1}{2\pi i}\int_{\mathfrak{C}'} \frac{dw}{w - w_1} = \frac{1}{2\pi i}\int_{\mathfrak{C}} \frac{f'(z)}{f(z) - w_1}\, dz,$$

i.e., w_1 is assumed by $f(z)$ precisely once inside $\mathfrak{C}$. $\mathfrak{G}'$ thus fills out the interior of $\mathfrak{C}'$ precisely once. (The still conceivable case that a z *inside* $\mathfrak{G}$ have its image w on the boundary $\mathfrak{C}'$ of $\mathfrak{G}$ is excluded by the theorem on the preservation of regions (B, § 1).)

4. These extensions of the preceding theorem are taken care of by means of suitable inversions. a) First, let $\mathfrak{C}'$ pass through ∞. Then $\mathfrak{C}'$ divides the plane into two regions, and $\mathfrak{G}'$ again lies in only one of the two. Let w_1 be a point in the interior of the other, and let $R > 0$ be chosen so that $\mathfrak{G}'$ and $\mathfrak{C}'$ lie wholly outside $|w - w_1| = R$. If we now set $\omega = \dfrac{R^2}{w - w_1}$, we obtain instead of $\mathfrak{C}'$ a curve Γ' which lies entirely in the finite part of the plane, to which the theorem of the preceding problem can now be applied. If z_0 was the point of $\mathfrak{C}$ which was sent to ∞, then $\dfrac{R^2}{f(z) - w_1}$ must now be regular and $= 0$ at z_0. This means, however, that $f(z)$ can only have a pole at z_0.

b) If $\mathfrak{C}$ also goes through ∞, a transformation of the form $\zeta = \dfrac{r^2}{z - z_1}$ leads back to the hitherto existing case. The assumption of regularity at the point ∞ lying on $\mathfrak{C}$ is thus to be understood according to K I, § 32. If ∞ now plays the role of the point z_0 used in a), then $f(z)$ may only have a pole at ∞.

5. Let $z_0, z_1, \ldots, z_n$ be a division of $\mathfrak{k}$ as in K I, § 8, let $f(z_\nu) = w_\nu$, so that $w_0, w_1, \ldots, w_n$ is a division of the continuous curve $\mathfrak{k}'$, and finally let M be an upper bound of $|f'(z)|$ along $\mathfrak{k}$. Then, since

$$|w_\nu - w_{\nu-1}| = \left| \int_{z_{\nu-1}}^{z_\nu} f'(z)\,dz \right| \leqq |z_\nu - z_{\nu-1}| \cdot M,$$

$\Sigma |w_\nu - w_{\nu-1}|$ is bounded as well as $\Sigma |z_\nu - z_{\nu-1}|$. It is easy to convince oneself that every division of $\mathfrak{k}'$ can be obtained in this way. Consequently, $\mathfrak{k}'$ is rectifiable, and is therefore a path. According to § 4, Prob. 2, one can set more precisely $w_\nu - w_{\nu-1} = (z_\nu - z_{\nu-1}) \cdot f'(\zeta'_\nu)$, where ζ'_ν denotes a point which approaches the point z_ν uniformly for all ν under refinement of the division. Hence, if ζ_ν denotes any point on the segment $z_{\nu-1} \ldots z_\nu$ of the path $\mathfrak{k}$, then

$$\lim_{n\to\infty} \sum_{\nu=1}^{n} |w_\nu - w_{\nu-1}| = \lim_{n\to\infty} \sum_{\nu=1}^{n} |z_\nu - z_{\nu-1}| \cdot |f'(\zeta_\nu)|,$$

and consequently

$$\mathfrak{k}' = \int_{\mathfrak{k}} |f'(z)| \cdot |dz|$$

in the sense of § 4, Prob. 3.

6. Let $f(z) = \sum_{n=0}^{\infty} a_n z^n$ in $|z| < r$. If the calculation given in B, § 19, pp. 86–87 for a series $z + a_2 z^2 + \ldots$ is carried out for the present case, we obtain quite analogously

$$\mathcal{J}(\rho) = (|a_1|^2\rho^2 + 2|a_2|^2\rho^4 + \cdots + n|a_n|^2\rho^{2n} + \cdots)\pi.$$

The meaning of the possible zeros of $f'(z)$ as regards the transformation determinant of the double integral (B, p. 87) is that it is 0 at (a finite number of) isolated points but is otherwise invariably >0. One can therefore divide the circle into a finite number of subdomains in such a manner that the mapping is simple (*schlicht*) for each individual part and therefore the transformation of the double integral is permissible according to the classical theorems of the integral calculus. Is the formula still valid, or to what extent is the formula still valid, for $\rho = r$?

7. The ratio is equal to $\dfrac{\mathcal{J}(\rho)}{\rho^2\pi}$ and consequently $\to |a_1|^2$. According

to the arguments given in connection with the preceding problem, this is also valid for $a_1 = 0$.

8. a) According to Prob. 5, it is equal to the integral $\int |f'(z)||dz|$ taken along $|z| = r$.

b) According to Prob. 6 and B, § 19, it is equal to the double integral $\int\int |f'(z)|^2\, dxdy$ extended over the disk $|z| \leqq r$ in the xy-plane.

c) $=$ angle of inclination of the tangent to the circle $+$ the rotation $= \text{am}\, z + \frac{\pi}{2} + \text{am}\, f'(z) = \text{am}\, (izf'(z))$.

d) Set $z = re^{i\varphi}$. By definition the curvature is equal to the derivative with respect to φ of the angle of inclination divided by the derivative with respect to φ of the arc length. According to c) the angle of inclination equals

$$\text{am}\, (izf'(z)) = \mathfrak{J}\, (\log izf'(z)).$$

Its derivative with respect to φ is

$$\mathfrak{J}\left(\frac{d}{dz}(\log izf'(z)) \cdot \frac{dz}{d\varphi}\right) = \mathfrak{J}\left(\frac{zf''(z) + f'(z)}{zf'(z)} \cdot iz\right) = \mathfrak{R}\left(1 + z\frac{f''(z)}{f'(z)}\right).$$

According to a), the derivative, with respect to φ, of the arc length is equal to

$$|f'(z)| \cdot |z|$$

(proof?). Hence, the desired curvature equals

$$\frac{\mathfrak{R}\left(1 + z\dfrac{f''(z)}{f'(z)}\right)}{|zf'(z)|}.$$

9. Since only the regularity for $|z| = r$ came into consideration in connection with questions a), c), and d), no changes at all are necessary here.

b) If $f(z) = a_0 + \frac{a_1}{z} + \frac{a_2}{z^2} + \cdots$ for $|z| \geqq r$, then the image of this region is identical with the image of $|z'| \leqq \frac{1}{r}$ by the function $\varphi(z') = a_0 + a_1 z' + a_2 z'^2 + \cdots$.

According to Prob. 6, therefore, the area of the image is equal to

$$\left(\frac{|a_1|^2}{r^2} + 2\frac{|a_2|^2}{r^4} + \cdots + n\frac{|a_n|^2}{r^{2n}} + \cdots\right)\pi.$$

10. Since $\varphi(\zeta)$ is supposed to effect a one-to-one mapping, $\varphi'(\zeta) \neq 0$ in $|\zeta| < 1$, and hence the inverse function $\Phi(w)$ is regular inside $\mathfrak{G}$ and its functional values there are < 1 in absolute value. Consequently $\Phi(f(z)) = F(z)$ is regular in $|z| < 1$, and $|F(z)| < 1$ there. Finally, $F(0) = \Phi(w_0) = 0$. Applying Schwarz's lemma to $F(z)$, we find that $|F(z)| \leqq |z|$ in $|z| < 1$, and that equality takes place only if $F(z)$ has the form $e^{i\alpha}z$ for a fixed α. Now if $|z| \leqq \rho < 1$, then also $|F(z)| \leqq |z| \leqq \rho$ there, and consequently $\varphi(F(z)) = f(z)$ lies in $\mathfrak{G}_\rho$, and, in fact, in its interior, unless $F(z) = e^{i\alpha}z$, i.e., $f(z) = \varphi(e^{i\alpha}z)$.

11. By means of $w = \varphi(z) = \dfrac{-\bar{w}_0 z + w_0}{z+1} = -u_0\dfrac{z-1}{z+1} + iv_0$, $|z| < 1$ is mapped on $\mathfrak{G}$, i.e., on $\Re(w) > 0$, in such a manner that 0 is transformed into w_0. The circle $|z| \leqq \rho$ thereby goes over into a circle $\bar{\mathfrak{G}}_\rho$ which has the segment $u_0\dfrac{1-\rho}{1+\rho} + iv_0 \ldots u_0\dfrac{1+\rho}{1-\rho} + iv_0$ for a diameter. Thus if $f(z)$ is a regular function in $|z| < 1$, whose values there lie in $\mathfrak{G}$, then they lie actually in $\bar{\mathfrak{G}}_\rho$ if $|z| \leqq \rho$. Because of the asserted position of $\bar{\mathfrak{G}}_\rho$, this already proves the assertion. The equality sign can only hold for $f(z) = \varphi(e^{i\alpha}z)$.

12. Since the radius of the circle $\bar{\mathfrak{G}}_\rho$ is equal to $\dfrac{2\rho u_0}{1-\rho^2}$, we have obviously

$$v_0 - \frac{2\rho u_0}{1-\rho^2} \leqq \Im f(z) \leqq v_0 + \frac{2\rho u_0}{1-\rho^2}$$

and

$$|f(z)| \leqq |u_0 + iv_0| + \frac{4\rho u_0}{1-\rho^2}.$$

13. Let $|z_1| < 1$, $|z_2| < 1$, and $z_1 \neq z_2$. Then

$$\frac{f(z_2) - f(z_1)}{z_2 - z_1} = a_1 + a_2(z_2 + z_1) + \cdots$$
$$+ a_n(z_2^{n-1} + z_2^{n-2} z_1 + \cdots + z_1^{n-1}) + \cdots.$$

Hence $\left|\frac{f(z_2)-f(z_1)}{z_2-z_1}\right| > |a_1| - 2|a_2| - 3|a_3| - \cdots \geqq 0$, and therefore $f(z_1) \neq f(z_2)$. The function $f(z)$ in § 18, Prob. 6 furnishes an example of this theorem.

14. If $\lim f_n(z) = f(z)$, then at any rate $f(z)$ is regular inside $\mathfrak{G}$. Let z_1 and z_2 lie in $\mathfrak{G}$ and let $z_1 \neq z_2$. The sequence of functions $f_n(z) - f_n(z_1)$ converges, and in fact uniformly in every subregion $\overline{\mathfrak{G}}'$, to the function $f(z) - f(z_1)$. According to the theorem in § 5, Prob. 4, this function has no other zeros in $\mathfrak{G}$ than the limit points there of zeros of $f_n(z) - f_n(z_1)$. These functions, however, vanish only for $z = z_1$, because the $f_n(z)$ are simple. Hence, $f(z_2) - f(z_1) \neq 0$.

15. If $f(z)$ is odd, then $(f(z))^2$ is an even function, and hence a regular function of $\zeta = z^2$ in $|\zeta| < 1$. Let $(f(z))^2 = \varphi(\zeta)$. This function is also simple. For if $|\zeta_1| < 1$, $|\zeta_2| < 1$, $\zeta_1 \neq \zeta_2$, then, if z_1 is a square root of ζ_1 and z_2 is a square root of ζ_2, we have also $z_1 \neq \pm z_2$, hence $f(z_1) \neq \pm f(z_2)$, and consequently $\varphi(\zeta_1) \neq \varphi(\zeta_2)$. If we now apply the inequality in B, p. 106 to $\varphi(\zeta)$, it follows that

$$\frac{|\zeta|}{(1+|\zeta|)^2} \leqq |\varphi(\zeta)| \leqq \frac{|\zeta|}{(1-|\zeta|)^2}$$

for $|\zeta| < 1$. If we reintroduce z and $f(z)$, we have the assertion.

16. If $w = f(z)$ and $w = \varphi(z)$ are two functions, both of which effect the mapping described in the problem, and if $z = \tilde{\varphi}(w)$ is the inverse of the function $w = \varphi(z)$, then $f(\tilde{\varphi}(w))$ maps a circle about 0 on another circle about 0 in such a manner that the center remains fixed. Therefore, according to B, § 5, 4,

$$f(\tilde{\varphi}(w)) = c \cdot w, \text{ that is, } f(z) = c \cdot \varphi(z).$$

Comparison of the derivatives at $z = a$ shows that $c = +1$. Thus the mappings are identical; the size of the image circle is uniquely determined.

§ 21. *Specific Mapping Porblems*

1. If we map $\mathfrak{K}_1$ and $\mathfrak{K}_2$ on concentric circles $\mathfrak{K}_1'$ and $\mathfrak{K}_2'$ according to I, § 12, Prob. 20, then the $\mathfrak{k}_\nu$ go over into congruent circles whose

disks lie in the ring between $\mathfrak{K}_1'$ and $\mathfrak{K}_2'$ and are tangent to these two circles. The chain now will close or not according as the constant angle subtended at the origin by these images of the $\mathfrak{k}_\nu$ bears a rational ratio to 2π or not. If the ratio mentioned is equal to p/q (p, q relatively prime integers), the chain closes with the circle $\mathfrak{k}_q$. As we see, the solution is considerably easier than the precise formulation of the problem.

2. First carry z_1 to 0 by means of a translation, and follow this by a rotation about 0 which brings the center of $\mathfrak{K}_2$ to the positive axis of reals; i.e., map the z-plane on an auxiliary plane by means of $\zeta = \dfrac{|z_2 - z_1|}{z_2 - z_1}(z - z_1)$. The points ζ_0 and ζ_0' which are conjugate simultaneously to the points of intersection of both circles with the real axis are now calculated to be $\zeta_0 = -\frac{2}{15}$ and $\zeta_0' = -\frac{32}{15}$. If these are carried to 0 and ∞ by means of $w = \dfrac{\zeta - \zeta_0}{\zeta - \zeta_0'}$, we obtain as images of $\mathfrak{K}_1$ and $\mathfrak{K}_2$ two circles about 0 with radii $R_1 = \frac{1}{4}$ and $R_2 = \frac{3}{4}$. Now $\dfrac{1}{\pi}\arcsin\dfrac{R_2 - R_1}{R_2 + R_1}$ must be a rational number $\dfrac{p}{q}$. Since we get $\dfrac{p}{q} = \frac{1}{6}$, the chain of the $\mathfrak{k}_\nu$ invariably closes with the sixth circle, no matter how we begin with $\mathfrak{k}_1$ and $\mathfrak{k}_2$.

3. Because of the group property of the linear functions (Elem., § 19), every $z^{(n)}$ is a linear function of z. These obviously have again the same fixed points for every n. The quadratic equations (Elem., § 20) $c_n z^2 + (d_n - a_n)z - b_n = 0$ therefore have the same roots as $cz^2 + (d - a)z - b = 0$ for every n. The assertion follows from this. If the mapping is brought into the fixed-point normal form according to I, § 12, Prob. 21, the detailed solution of this problem (see also Elem., § 20) enables one to see immediately the effect of an n-fold iteration of the same mapping.

4. We consider a $\xi\eta\zeta$-coordinate system in space such that the ξ- and η-axes coincide with the x-and y- axes, respectively, in the z-plane and such that the ζ-axis is perpendicular to the z-plane. Then the equation of the number sphere is $\xi^2 + \eta^2 + \zeta(\zeta - 1) = 0$, and the

parametric equations of the straight line containing the north pole and the point (ξ_0, η_0, ζ_0) on the sphere are $\xi = \xi_0(1 + t)$, $\eta = \eta_0 (1 + t)$, $\zeta = \zeta_0 + (\zeta_0 - 1)t$. For $t = \zeta_0/(1 - \zeta_0)$ we get $\zeta = 0$, and hence, if we suppress the subscript 0,

$$x = \frac{\xi}{1 - \zeta}, \qquad y = \frac{\eta}{1 - \zeta},$$

$$dx = \frac{(1 - \zeta)d\xi + \xi d\zeta}{(1 - \zeta)^2}, \qquad dy = \frac{(1 - \zeta)d\eta + \eta d\zeta}{(1 - \zeta)^2}.$$

The conformality of the mapping now requires that the line elements in the two surfaces, emanating from the point $z = x + iy$ in the z-plane and from the point (ξ, η, ζ) on the sphere, stand in a ratio to one another which depends only on the position of the pair of points and not on the direction of the line element. If, however, we take into consideration the equation of the sphere, we have

$$dx^2 + dy^2 = (d\xi^2 + d\eta^2 + d\zeta^2)/(1 - \zeta)^2,$$

which contains the proof of the assertion.

5. If we use λ and β as parameters, and if we leave out the common factor R throughout, then the equations of the sphere are

$$\xi = \cos\beta \cos\lambda, \quad \eta = \cos\beta \sin\lambda, \quad \zeta = \sin\beta.$$

Hence,

$$d\xi^2 + d\eta^2 + d\zeta^2 = d\beta^2 + \cos^2\beta d\lambda^2.$$

And we obtain quite directly

$$dx^2 + dy^2 = d\lambda^2 + \frac{d\beta^2}{\cos^2\beta} = \frac{d\xi^2 + d\eta^2 + d\zeta^2}{\cos^2\beta},$$

which proves the conformality.

6. By means of $z' = 4z + 1$, the beginning of the cut is moved to 0 and the original origin is sent to $+1$. $z'' = \sqrt{z'} = \sqrt{4z + 1}$ transforms the new region into the right half-plane, if we take the principal value of the square root. This half-plane is mapped by $w = \dfrac{z'' - 1}{z'' + 1}$ on the unit circle in the w-plane. Thus the function

$$w = \frac{\sqrt{4z+1} - 1}{\sqrt{4z+1} + 1} \quad \text{or} \quad z = \frac{w}{(1-w)^2}$$

does what is required, since, moreover, it vanishes at $z = 0$ and its derivative there is equal to 1. This is the mapping which plays an exceptional role in connection with many problems (B, § 22).

7. The equation can be written in the form

$$a\left(w + \frac{b}{2a}\right)^2 = z + \frac{b^2 - 4ac}{4a}.$$

Hence, if we set

$$\sqrt{a}\left(w + \frac{b}{2a}\right) = w', \quad z + \frac{b^2 - 4ac}{4a} = z',$$

we have to investigate simply $w'^2 = z'$, $w' = \sqrt{z'}$. We have thus a two-sheeted surface with branch-points at $-\frac{b^2 - 4ac}{4a}$ and ∞. If z goes into the first of these points, then w goes to $-\frac{b}{2a}$; if z encircles it once, w' goes into $-w'$, i.e., w goes over into $-w - \frac{b}{a}$, or, in other words, into that point which is symmetric to w with respect to the point $-\frac{b}{2a}$.

8. By means of $z' = \frac{z+2}{z-2}$ the cut is carried into the negative axis of reals, $z'' = \sqrt{z'}$ transforms the resulting slit plane into the right half-plane, and the latter is mapped on the unit circle by $w = \frac{z''-1}{z''+1}$. Hence,

$$w = \frac{\sqrt{\frac{z+2}{z-2}} - 1}{\sqrt{\frac{z+2}{z-2}} + 1} \quad \text{or} \quad z = w + \frac{1}{w}.$$

Here the principal value of the square root is to be taken; with the other value of the square root the mapping of the given domain on the exterior of the unit circle is effected. (Cf. B, § 12.)

9. If $z' = x' + iy'$ is the reflection of $z = x + iy$ with respect to the unit circle, then

$$x' = \frac{x}{x^2 + y^2} = \frac{\cos t}{2a(1 - \cos t)}, \quad y' = \frac{y}{x^2 + y^2} = \frac{\sin t}{2a(1 - \cos t)}$$

or

$$x' + \frac{1}{4a} = \frac{1}{4a} \cot^2 \frac{t}{2}, \quad y'^2 = \frac{1}{4a^2} \cot^2 \frac{t}{2}$$

or finally $y'^2 = \frac{1}{a}\left(x' + \frac{1}{4a}\right)$ furnishes the reflection of the cardioid. This is a parabola with focus at 0 and vertex at $-\frac{1}{4a}$. To the exterior of the cardioid corresponds the interior of the parabola which — if we again suppress the prime marks — can be defined by $\mathfrak{J}(\sqrt{z}) < \frac{1}{2\sqrt{a}}$, $\sqrt{a} > 0$. According to I, § 13, Prob. 14, $\tan^2\left(\frac{\pi\sqrt{a}}{2i}\sqrt{z}\right)$ maps this parabola interior on the interior of the unit circle. Hence, the exterior of the cardioid in the z-plane and the interior of the unit circle in the w-plane are mapped on one another by means of

$$w = \tan^2\left(\frac{\pi\sqrt{a}}{2i}\sqrt{\frac{1}{z}}\right) \quad \text{or } z = \frac{a\pi^2}{\log^2\left(\frac{i + \sqrt{w}}{i - \sqrt{w}}\right)}.$$

10. If $\omega = 2\alpha$ and $-i\omega' = 2\beta$ are positive, then, according to § 14, Prob. 7, $\wp\,(z|\alpha, i\beta)$ maps the rectangle $0 \leqq x \leqq \alpha$, $0 \leqq y \leqq \beta$ on the lower half-plane. Therefore, if we take $\alpha = 4$, $\beta = 2$, then $\wp\,(z + 2 + i|4, 2i)$ maps the rectangle presented in the problem on the lower half-plane. Hence, this rectangle is mapped on the unit circle in the w-plane by means of

$$w = \frac{\wp(z + 2 + i \mid 4, 2i) + i}{\wp(z + 2 + i \mid 4, 2i) - i}.$$

11. We shall show that the upper half of the z-plane is mapped on

an equilateral triangle in the w-plane by means of the integral $w = \int_0^z t^{-\frac{2}{3}}(1-t)^{-\frac{2}{3}}dt$ for a suitable choice of the determination of the integrand. To this end we have merely to determine am t and am $(1-t)$ so that they remain single-valued and continuous provided that $\mathfrak{J}(t)$ remains $\geqq 0$ (at $t = 0$ and $+1$ naturally only as limiting values). This is obviously the case if we agree that we shall always take

$$0 \leqq \text{am}\, t \leqq +\pi \quad \text{and} \quad -\pi \leqq \text{am}\,(1-t) \leqq 0.$$

Now first if $z = x$, $0 < x < 1$, then

$$w = \int_0^x t^{-\frac{2}{3}}(1-t)^{-\frac{2}{3}}\,dt$$

taken rectilinearly. Thus, as x increases from 0 to 1, w moves monotonically to the right from 0 to the point

$$w = s = \int_0^1 t^{-\frac{2}{3}}(1-t)^{-\frac{2}{3}}\,dt.$$

[The value of this Eulerian integral of the first kind, as is well known, can be given in closed form in terms of the Γ-function. It is $s = (\Gamma(\frac{1}{3}))^2/\Gamma(\frac{2}{3})$.] If now $z = x$, $1 < x < +\infty$, then, since the integrals remain continuous at 0 and 1, we set

$$w = s + \int_1^x t^{-\frac{2}{3}}(1-t)^{-\frac{2}{3}}\,dt.$$

Now, however, $\text{am}(1-t) = -\pi$, and hence $(1-t)^{-\frac{2}{3}} = e^{+\frac{2}{3}\pi i}(t-1)^{-\frac{2}{3}}$, so that we obtain

$$w = s + e^{\frac{2\pi i}{3}} \int_1^x t^{-\frac{2}{3}}(t-1)^{-\frac{2}{3}}\,dt.$$

The new integral is again positive and, as x proceeds from $+1$ to $+\infty$, increases monotonically from 0 to the value $\int_1^\infty t^{-\frac{2}{3}}(t-1)^{-\frac{2}{3}}dt$, which the substitution $t = \frac{1}{u}$ shows to be equal to s once more. Thus,

w starts at $+s$ and moves the distance s along the ray whose angle of inclination is $\frac{2}{3}\pi$, arriving at the point $s(1 + e^{\frac{2}{3}\pi i}) = se^{\frac{\pi i}{3}}$. Finally, if $z = -x$ and $x > 0$, then along the straight path from 0 to z we now have to set am $t = \pi$, am $(1 - t) = 0$. If we also put $|t| = \tau$, we get

$$w = -e^{-\frac{2}{3}\pi i} \int_0^x \tau^{-\frac{2}{3}} (1 + \tau)^{-\frac{2}{3}} \, d\tau.$$

If x goes from 0 to $+\infty$, this new integral is also positive and increases monotonically from 0 to the value $\int_0^\infty \tau^{-\frac{2}{3}} (1 + \tau)^{-\frac{2}{3}} \, d\tau$, which the substitution $\tau = \frac{t}{1 - t}$ shows is again the value s. Thus, if z runs from 0 to $-\infty$, w moves the distance s from 0 in the direction of $-e^{-\frac{2\pi i}{3}} = e^{\frac{\pi i}{3}}$, arriving once more at the point $se^{\frac{\pi i}{3}}$.

We see then that if z describes the entire real axis in the positive sense, w describes the equilateral triangle with the vertices 0, s, $se^{\frac{\pi i}{3}}$ in the positive sense. Hence,

$$w = \frac{1}{s} \int_0^z t^{-\frac{2}{3}} (1 - t)^{-\frac{2}{3}} \, dt = f(z)$$

describes the triangle presented in the problem. The single-valued function $f(z)$ defined by this integral in the closed upper half-plane therefore maps this half-plane on the given triangle in a one-to-one and, except at the points 0, 1, ∞, conformal manner. The inverse function consequently furnishes the mapping required in the problem.

12. a) The multiple-valuedness comes about through the poles of the integrand at 1, ω, ω^2 $\left(\text{where } \omega = e^{\frac{2\pi i}{3}}\right)$, at which there are the residues $\frac{1}{3}$, $\frac{\omega^2}{3}$, and $\frac{\omega}{3}$. In the finite part of the plane the derivative $f_1'(z)$ vanishes only at 0 and to the first order; the neighborhoods of the origin are therefore not mapped simply; angles there are doubled. At ∞ the integrand vanishes to the second order, so that $f_1(z)$ is single-

valued in a neighborhood of ∞ and regular at ∞ itself (cf. § 7, Prob. 8). We therefore cut the plane only along the three segments from 0 to 1, from 0 to ω, and from 0 to ω^2. We now let z describe the six banks of the cuts in such a manner that the plane lies to the left. If we allow 1) z to run from 0 to 1 along the upper bank of the cut, we may let w begin with 0; it then proceeds to $-\infty$ along the negative axis of reals. Since the opposite points of the same cut are reached by making a single negative circuit of the pole $+1$ having the residue $\frac{1}{3}$, each of the values w there is greater by $-2\pi i\cdot\frac{1}{3}$. Hence, if 2) z goes from $+1$ back to 0 along the lower bank, w retraces the negative axis of reals, displaced by $-\frac{2\pi i}{3}$, up to the point $-\frac{2\pi i}{3}$. If now 3) z goes from 0 to ω^2, the path of z has the angle $-\frac{2}{3}\pi$ at 0; that of w must therefore have the angle $-\frac{4}{3}\pi$ at $-\frac{2\pi i}{3}$. Hence, w now proceeds in the direction $-\frac{\pi}{3}$ and again rectilinearly (why?) to ∞. If 4) z returns to 0 after a negative encirclement of the pole ω^2 having the residue $\frac{\omega}{3}$, w runs through the same values as before, in reverse order and increased by $-2\pi i\frac{\omega}{3}$; and thus describes the line which is parallel to the last line and contains the point $-2\pi i\cdot\frac{1}{3} - 2\pi i\frac{\omega}{3} = 2\pi i\cdot\frac{\omega^2}{3}$, up to this point. If now 5) z goes from 0 to ω, the path of z again has an angle of $-\frac{2}{3}\pi$, that of ω therefore one of $-\frac{4}{3}\pi$. Consequently, w proceeds from the last point in the direction $+\frac{\pi}{3}$ to ∞ once more. If z circuits the pole ω, having the residue $\frac{\omega^2}{3}$, in the negative sense, and 6) returns to 0, w returns with the values increased by $-2\pi i\cdot\frac{\omega^2}{3}$, and hence parallel to the preceding direction and up to the point $2\pi i\frac{\omega^2}{3} - 2\pi i\frac{\omega^2}{3} = 0$. The path taken by w is closed. Our slit z-plane is thus mapped by $w = f_1(z)$ in a one-to-one and, except for the points 0, 1, ω, ω^2, conformal manner on the region to the left of the described w-path. This region consists of the equilateral triangle with vertices 0, $-2\pi i\cdot\frac{1}{3}$, and three half-strips affixed to its sides perpendicularly and in the outward direction.

b) $f_2(z)$ can be reduced to $f_1(z)$ by means of the substitution $z = \frac{1}{z'}$. We have $f_2(z) = f_1\left(\frac{1}{z}\right) + $ a constant.

13. If z lies in $\mathfrak{G}$, then $z' = \frac{1}{2}\pi z^2$ lies in the strip $0 < \mathfrak{I}(z') < \pi$ and hence $z'' = e^{z'} = e^{\frac{1}{2}\pi z^2}$ lies in the upper half of the z''-plane. The value a first becomes $a' = \frac{1}{2}\pi a^2$ and then $a'' = e^{a'} = e^{\frac{1}{2}\pi a^2}$, which is a positive imaginary. Through $z''' = \dfrac{z'' - a''}{z'' - \bar{a}''}$ we obtain the unit circle in the z'''-plane; a'' is transformed into 0. We have

$$\frac{dz'''}{dz} = \frac{dz'''}{dz''} \cdot \frac{dz''}{dz} = \frac{a'' - \bar{a}''}{(z'' - \bar{a}'')^2} \cdot z'' \cdot \pi z.$$

For $z = a$ this is equal to $\frac{1}{2}\pi a$ because $\bar{a}'' = -a''$. Hence, $w = \dfrac{2}{\pi a} \dfrac{e^{\frac{1}{2}\pi z^2} - e^{\frac{1}{2}\pi a^2}}{e^{\frac{1}{2}\pi z^2} - e^{\frac{1}{2}\pi \bar{a}^2}}$ maps the given region on a circle about $w = 0$ in such a manner that a is transformed into 0 and the derivative there is equal to 1. The radius of this circle is $r(\mathfrak{G}; a) = \dfrac{2\cdot}{\pi |a|}$.

14. a) If $|a - z_0| = \rho < R$, then

$$w = (R^2 - \rho^2) \frac{z - a}{R^2 - (\bar{a} - \bar{z}_0)(z - z_0)}$$

furnishes the required mapping; $z = z_0 + R$ gives $r(\mathfrak{G}; a) = \dfrac{R^2 - \rho^2}{R}$. The surface is thus a paraboloid of revolution which passes through the circumference $|z - z_0| = R$ and whose vertex lies above z_0.

b) If $\mathfrak{I}(a) > 0$, then $w = (a - \bar{a}) \dfrac{z - a}{z - \bar{a}}$ furnishes the required mapping; $z = 0$ gives $r(\mathfrak{G}; a) = |a - \bar{a}| = 2\mathfrak{I}(a)$. The surface is thus a plane which intersects the z-plane along the real axis.

c) If $0 < \mathfrak{I}(a) < \pi$, then (cf. Prob. 13).

$$w = e^{-a} (e^a - e^{\bar{a}}) \frac{e^z - e^a}{e^z - e^{\bar{a}}}$$

furnishes the required mapping; $z = 0$ gives $r(\mathfrak{G}; a) = |1 - e^{\bar{a} - a}| = 2 \sin \beta$ if we set $a = \alpha + i\beta$. The surface is a cylindrical surface whose generators are parallel to the real axis.

d) The calculations in c) and in Prob. 13 show that the required mapping in this case is furnished by

$$w = \frac{1}{\pi a}\left(1 - e^{\frac{\pi}{2}(\bar{a}^2 - a^2)}\right) \frac{e^{\frac{1}{2}\pi z^2} - e^{\frac{1}{2}\pi a^2}}{e^{\frac{1}{2}\pi z^2} - e^{\frac{1}{2}\pi \bar{a}^2}}.$$

For $z = 0$ we find that

$$r(\mathfrak{G}; a) \quad \frac{1}{\pi |a|} \cdot \left| 1 - e^{\frac{\pi}{2}(\bar{a}^2 - a^2)} \right| = \frac{2 \sin \alpha\beta\pi}{\pi \sqrt{\alpha^2 + \beta^2}}$$

if we set $a = \alpha + i\beta$. The equation of the surface is hereby given in terms of the variables α, β.

Note that in each of the four examples, the ordinate to the surface in question has the limit 0 as a approaches the boundary of $\mathfrak{G}$.

A CATALOGUE OF SELECTED DOVER BOOKS
IN ALL FIELDS OF INTEREST

A CATALOGUE OF SELECTED DOVER BOOKS IN ALL FIELDS OF INTEREST

CONDITIONED REFLEXES, Ivan P. Pavlov. Full translation of most complete statement of Pavlov's work; cerebral damage, conditioned reflex, experiments with dogs, sleep, similar topics of great importance. 430pp. 5⅜ x 8½. 60614-7 Pa. $4.50

NOTES ON NURSING: WHAT IT IS, AND WHAT IT IS NOT, Florence Nightingale. Outspoken writings by founder of modern nursing. When first published (1860) it played an important role in much needed revolution in nursing. Still stimulating. 140pp. 5⅜ x 8½. 22340-X Pa. $3.00

HARTER'S PICTURE ARCHIVE FOR COLLAGE AND ILLUSTRATION, Jim Harter. Over 300 authentic, rare 19th-century engravings selected by noted collagist for artists, designers, decoupeurs, etc. Machines, people, animals, etc., printed one side of page. 25 scene plates for backgrounds. 6 collages by Harter, Satty, Singer, Evans. Introduction. 192pp. 8⅞ x 11¾. 23659-5 Pa. $5.00

MANUAL OF TRADITIONAL WOOD CARVING, edited by Paul N. Hasluck. Possibly the best book in English on the craft of wood carving. Practical instructions, along with 1,146 working drawings and photographic illustrations. Formerly titled *Cassell's Wood Carving*. 576pp. 6½ x 9¼. 23489-4 Pa. $7.95

THE PRINCIPLES AND PRACTICE OF HAND OR SIMPLE TURNING, John Jacob Holtzapffel. Full coverage of basic lathe techniques—history and development, special apparatus, softwood turning, hardwood turning, metal turning. Many projects—billiard ball, works formed within a sphere, egg cups, ash trays, vases, jardiniers, others—included. 1881 edition. 800 illustrations. 592pp. 6⅛ x 9¼. 23365-0 Clothbd. $15.00

THE JOY OF HANDWEAVING, Osma Tod. Only book you need for hand weaving. Fundamentals, threads, weaves, plus numerous projects for small board-loom, two-harness, tapestry, laid-in, four-harness weaving and more. Over 160 illustrations. 2nd revised edition. 352pp. 6½ x 9¼. 23458-4 Pa. $6.00

THE BOOK OF WOOD CARVING, Charles Marshall Sayers. Still finest book for beginning student in wood sculpture. Noted teacher, craftsman discusses fundamentals, technique; gives 34 designs, over 34 projects for panels, bookends, mirrors, etc. "Absolutely first-rate"—E. J. Tangerman. 33 photos. 118pp. 7¾ x 10⅝. 23654-4 Pa. $3.50

HOLLYWOOD GLAMOUR PORTRAITS, edited by John Kobal. 145 photos capture the stars from 1926-49, the high point in portrait photography. Gable, Harlow, Bogart, Bacall, Hedy Lamarr, Marlene Dietrich, Robert Montgomery, Marlon Brando, Veronica Lake; 94 stars in all. Full background on photographers, technical aspects, much more. Total of 160pp. 8⅜ x 11¼. **23352-9 Pa. $6.00**

THE NEW YORK STAGE: FAMOUS PRODUCTIONS IN PHOTOGRAPHS, edited by Stanley Appelbaum. 148 photographs from Museum of City of New York show 142 plays, 1883-1939. *Peter Pan, The Front Page, Dead End, Our Town,* O'Neill, hundreds of actors and actresses, etc. Full indexes. 154pp. 9½ x 10. **23241-7 Pa. $6.00**

DIALOGUES CONCERNING TWO NEW SCIENCES, Galileo Galilei. Encompassing 30 years of experiment and thought, these dialogues deal with geometric demonstrations of fracture of solid bodies, cohesion, leverage, speed of light and sound, pendulums, falling bodies, accelerated motion, etc. 300pp. 5⅜ x 8½. **60099-8 Pa. $4.00**

THE GREAT OPERA STARS IN HISTORIC PHOTOGRAPHS, edited by James Camner. 343 portraits from the 1850s to the 1940s: Tamburini, Mario, Caliapin, Jeritza, Melchior, Melba, Patti, Pinza, Schipa, Caruso, Farrar, Steber, Gobbi, and many more—270 performers in all. Index. 199pp. 8⅜ x 11¼. **23575-0 Pa. $7.50**

J. S. BACH, Albert Schweitzer. Great full-length study of Bach, life, background to music, music, by foremost modern scholar. Ernest Newman translation. 650 musical examples. Total of 928pp. 5⅜ x 8½. (Available in U.S. only) **21631-4, 21632-2 Pa., Two-vol. set $11.00**

COMPLETE PIANO SONATAS, Ludwig van Beethoven. All sonatas in the fine Schenker edition, with fingering, analytical material. One of best modern editions. Total of 615pp. 9 x 12. (Available in U.S. only) **23134-8, 23135-6 Pa., Two-vol. set $15.50**

KEYBOARD MUSIC, J. S. Bach. Bach-Gesellschaft edition. For harpsichord, piano, other keyboard instruments. English Suites, French Suites, Six Partitas, Goldberg Variations, Two-Part Inventions, Three-Part Sinfonias. 312pp. 8⅛ x 11. (Available in U.S. only) **22360-4 Pa. $6.95**

FOUR SYMPHONIES IN FULL SCORE, Franz Schubert. Schubert's four most popular symphonies: No. 4 in C Minor ("Tragic"); No. 5 in B-flat Major; No. 8 in B Minor ("Unfinished"); No. 9 in C Major ("Great"). Breitkopf & Hartel edition. Study score. 261pp. 9⅜ x 12¼. **23681-1 Pa. $6.50**

THE AUTHENTIC GILBERT & SULLIVAN SONGBOOK, W. S. Gilbert, A. S. Sullivan. Largest selection available; 92 songs, uncut, original keys, in piano rendering approved by Sullivan. Favorites and lesser-known fine numbers. Edited with plot synopses by James Spero. 3 illustrations. 399pp. 9 x 12. **23482-7 Pa. $9.95**

THE DEPRESSION YEARS AS PHOTOGRAPHED BY ARTHUR ROTHSTEIN, Arthur Rothstein. First collection devoted entirely to the work of outstanding 1930s photographer: famous dust storm photo, ragged children, unemployed, etc. 120 photographs. Captions. 119pp. 9¼ x 10¾.
23590-4 Pa. $5.00

CAMERA WORK: A PICTORIAL GUIDE, Alfred Stieglitz. All 559 illustrations and plates from the most important periodical in the history of art photography, *Camera Work* (1903-17). Presented four to a page, reduced in size but still clear, in strict chronological order, with complete captions. Three indexes. Glossary. Bibliography. 176pp. 8⅜ x 11¼.
23591-2 Pa. $6.95

ALVIN LANGDON COBURN, PHOTOGRAPHER, Alvin L. Coburn. Revealing autobiography by one of greatest photographers of 20th century gives insider's version of Photo-Secession, plus comments on his own work. 77 photographs by Coburn. Edited by Helmut and Alison Gernsheim. 160pp. 8⅛ x 11. 23685-4 Pa. $6.00

NEW YORK IN THE FORTIES, Andreas Feininger. 162 brilliant photographs by the well-known photographer, formerly with *Life* magazine, show commuters, shoppers, Times Square at night, Harlem nightclub, Lower East Side, etc. Introduction and full captions by John von Hartz. 181pp. 9¼ x 10¾. 23585-8 Pa. $6.95

GREAT NEWS PHOTOS AND THE STORIES BEHIND THEM, John Faber. Dramatic volume of 140 great news photos, 1855 through 1976, and revealing stories behind them, with both historical and technical information. Hindenburg disaster, shooting of Oswald, nomination of Jimmy Carter, etc. 160pp. 8¼ x 11. 23667-6 Pa. $5.00

THE ART OF THE CINEMATOGRAPHER, Leonard Maltin. Survey of American cinematography history and anecdotal interviews with 5 masters—Arthur Miller, Hal Mohr, Hal Rosson, Lucien Ballard, and Conrad Hall. Very large selection of behind-the-scenes production photos. 105 photographs. Filmographies. Index. Originally *Behind the Camera.* 144pp. 8¼ x 11. 23686-2 Pa. $5.00

DESIGNS FOR THE THREE-CORNERED HAT (LE TRICORNE), Pablo Picasso. 32 fabulously rare drawings—including 31 color illustrations of costumes and accessories—for 1919 production of famous ballet. Edited by Parmenia Migel, who has written new introduction. 48pp. 9⅜ x 12¼. (Available in U.S. only) 23709-5 Pa. $5.00

NOTES OF A FILM DIRECTOR, Sergei Eisenstein. Greatest Russian filmmaker explains montage, making of *Alexander Nevsky,* aesthetics; comments on self, associates, great rivals (Chaplin), similar material. 78 illustrations. 240pp. 5⅜ x 8½. 22392-2 Pa. $4.50

CATALOGUE OF DOVER BOOKS

ART FORMS IN NATURE, Ernst Haeckel. Multitude of strangely beautiful natural forms: Radiolaria, Foraminifera, jellyfishes, fungi, turtles, bats, etc. All 100 plates of the 19th-century evolutionist's *Kunstformen der Natur* (1904). 100pp. 9⅜ x 12¼. 22987-4 Pa. $5.00

CHILDREN: A PICTORIAL ARCHIVE FROM NINETEENTH-CENTURY SOURCES, edited by Carol Belanger Grafton. 242 rare, copyright-free wood engravings for artists and designers. Widest such selection available. All illustrations in line. 119pp. 8⅜ x 11¼. 23694-3 Pa. $4.00

WOMEN: A PICTORIAL ARCHIVE FROM NINETEENTH-CENTURY SOURCES, edited by Jim Harter. 391 copyright-free wood engravings for artists and designers selected from rare periodicals. Most extensive such collection available. All illustrations in line. 128pp. 9 x 12. 23703-6 Pa. $4.50

ARABIC ART IN COLOR, Prisse d'Avennes. From the greatest ornamentalists of all time—50 plates in color, rarely seen outside the Near East, rich in suggestion and stimulus. Includes 4 plates on covers. 46pp. 9⅜ x 12¼. 23658-7 Pa. $6.00

AUTHENTIC ALGERIAN CARPET DESIGNS AND MOTIFS, edited by June Beveridge. Algerian carpets are world famous. Dozens of geometrical motifs are charted on grids, color-coded, for weavers, needleworkers, craftsmen, designers. 53 illustrations plus 4 in color. 48pp. 8¼ x 11. (Available in U.S. only) 23650-1 Pa. $1.75

DICTIONARY OF AMERICAN PORTRAITS, edited by Hayward and Blanche Cirker. 4000 important Americans, earliest times to 1905, mostly in clear line. Politicians, writers, soldiers, scientists, inventors, industrialists, Indians, Blacks, women, outlaws, etc. Identificatory information. 756pp. 9¼ x 12¾. 21823-6 Clothbd. $40.00

HOW THE OTHER HALF LIVES, Jacob A. Riis. Journalistic record of filth, degradation, upward drive in New York immigrant slums, shops, around 1900. New edition includes 100 original Riis photos, monuments of early photography. 233pp. 10 x 7⅞. 22012-5 Pa. $7.00

NEW YORK IN THE THIRTIES, Berenice Abbott. Noted photographer's fascinating study of city shows new buildings that have become famous and old sights that have disappeared forever. Insightful commentary. 97 photographs. 97pp. 11⅜ x 10. 22967-X Pa. $5.00

MEN AT WORK, Lewis W. Hine. Famous photographic studies of construction workers, railroad men, factory workers and coal miners. New supplement of 18 photos on Empire State building construction. New introduction by Jonathan L. Doherty. Total of 69 photos. 63pp. 8 x 10¾. 23475-4 Pa. $3.00

THE ANATOMY OF THE HORSE, George Stubbs. Often considered the great masterpiece of animal anatomy. Full reproduction of 1766 edition, plus prospectus; original text and modernized text. 36 plates. Introduction by Eleanor Garvey. 121pp. 11 x 14¾. 23402-9 Pa. $6.00

BRIDGMAN'S LIFE DRAWING, George B. Bridgman. More than 500 illustrative drawings and text teach you to abstract the body into its major masses, use light and shade, proportion; as well as specific areas of anatomy, of which Bridgman is master. 192pp. 6½ x 9¼. (Available in U.S. only) 22710-3 Pa. $3.50

ART NOUVEAU DESIGNS IN COLOR, Alphonse Mucha, Maurice Verneuil, Georges Auriol. Full-color reproduction of *Combinaisons ornementales* (c. 1900) by Art Nouveau masters. Floral, animal, geometric, interlacings, swashes—borders, frames, spots—all incredibly beautiful. 60 plates, hundreds of designs. 9⅜ x 8-1/16. 22885-1 Pa. $4.00

FULL-COLOR FLORAL DESIGNS IN THE ART NOUVEAU STYLE, E. A. Seguy. 166 motifs, on 40 plates, from *Les fleurs et leurs applications decoratives* (1902): borders, circular designs, repeats, allovers, "spots." All in authentic Art Nouveau colors. 48pp. 9⅜ x 12¼. 23439-8 Pa. $5.00

A DIDEROT PICTORIAL ENCYCLOPEDIA OF TRADES AND INDUSTRY, edited by Charles C. Gillispie. 485 most interesting plates from the great French Encyclopedia of the 18th century show hundreds of working figures, artifacts, process, land and cityscapes; glassmaking, papermaking, metal extraction, construction, weaving, making furniture, clothing, wigs, dozens of other activities. Plates fully explained. 920pp. 9 x 12. 22284-5, 22285-3 Clothbd., Two-vol. set $40.00

HANDBOOK OF EARLY ADVERTISING ART, Clarence P. Hornung. Largest collection of copyright-free early and antique advertising art ever compiled. Over 6,000 illustrations, from Franklin's time to the 1890's for special effects, novelty. Valuable source, almost inexhaustible.
Pictorial Volume. Agriculture, the zodiac, animals, autos, birds, Christmas, fire engines, flowers, trees, musical instruments, ships, games and sports, much more. Arranged by subject matter and use. 237 plates. 288pp. 9 x 12. 20122-8 Clothbd. $14.50

Typographical Volume. Roman and Gothic faces ranging from 10 point to 300 point, "Barnum," German and Old English faces, script, logotypes, scrolls and flourishes, 1115 ornamental initials, 67 complete alphabets, more. 310 plates. 320pp. 9 x 12. 20123-6 Clothbd. $15.00

CALLIGRAPHY (CALLIGRAPHIA LATINA), J. G. Schwandner. High point of 18th-century ornamental calligraphy. Very ornate initials, scrolls, borders, cherubs, birds, lettered examples. 172pp. 9 x 13. 20475-8 Pa. $7.00

YUCATAN BEFORE AND AFTER THE CONQUEST, Diego de Landa. First English translation of basic book in Maya studies, the only significant account of Yucatan written in the early post-Conquest era. Translated by distinguished Maya scholar William Gates. Appendices, introduction, 4 maps and over 120 illustrations added by translator. 162pp. 5⅜ x 8½.
23622-6 Pa. $3.00

THE MALAY ARCHIPELAGO, Alfred R. Wallace. Spirited travel account by one of founders of modern biology. Touches on zoology, botany, ethnography, geography, and geology. 62 illustrations, maps. 515pp. 5⅜ x 8½.
20187-2 Pa. $6.95

THE DISCOVERY OF THE TOMB OF TUTANKHAMEN, Howard Carter, A. C. Mace. Accompany Carter in the thrill of discovery, as ruined passage suddenly reveals unique, untouched, fabulously rich tomb. Fascinating account, with 106 illustrations. New introduction by J. M. White. Total of 382pp. 5⅜ x 8½. (Available in U.S. only) 23500-9 Pa. $4.00

THE WORLD'S GREATEST SPEECHES, edited by Lewis Copeland and Lawrence W. Lamm. Vast collection of 278 speeches from Greeks up to present. Powerful and effective models; unique look at history. Revised to 1970. Indices. 842pp. 5⅜ x 8½. 20468-5 Pa. $8.95

THE 100 GREATEST ADVERTISEMENTS, Julian Watkins. The priceless ingredient; His master's voice; 99 44/100% pure; over 100 others. How they were written, their impact, etc. Remarkable record. 130 illustrations. 233pp. 7⅞ x 10 3/5. 20540-1 Pa. $5.95

CRUICKSHANK PRINTS FOR HAND COLORING, George Cruickshank. 18 illustrations, one side of a page, on fine-quality paper suitable for watercolors. Caricatures of people in society (c. 1820) full of trenchant wit. Very large format. 32pp. 11 x 16. 23684-6 Pa. $5.00

THIRTY-TWO COLOR POSTCARDS OF TWENTIETH-CENTURY AMERICAN ART, Whitney Museum of American Art. Reproduced in full color in postcard form are 31 art works and one shot of the museum. Calder, Hopper, Rauschenberg, others. Detachable. 16pp. 8¼ x 11.
23629-3 Pa. $3.00

MUSIC OF THE SPHERES: THE MATERIAL UNIVERSE FROM ATOM TO QUASAR SIMPLY EXPLAINED, Guy Murchie. Planets, stars, geology, atoms, radiation, relativity, quantum theory, light, antimatter, similar topics. 319 figures. 664pp. 5⅜ x 8½.
21809-0, 21810-4 Pa., Two-vol. set $11.00

EINSTEIN'S THEORY OF RELATIVITY, Max Born. Finest semi-technical account; covers Einstein, Lorentz, Minkowski, and others, with much detail, much explanation of ideas and math not readily available elsewhere on this level. For student, non-specialist. 376pp. 5⅜ x 8½.
60769-0 Pa. $4.50

AMERICAN BIRD ENGRAVINGS, Alexander Wilson et al. All 76 plates. from Wilson's *American Ornithology* (1808-14), most important ornithological work before Audubon, plus 27 plates from the supplement (1825-33) by Charles Bonaparte. Over 250 birds portrayed. 8 plates also reproduced in full color. 111pp. 9⅜ x 12½. 23195-X Pa. $6.00

CRUICKSHANK'S PHOTOGRAPHS OF BIRDS OF AMERICA, Allan D. Cruickshank. Great ornithologist, photographer presents 177 closeups, groupings, panoramas, flightings, etc., of about 150 different birds. Expanded *Wings in the Wilderness.* Introduction by Helen G. Cruickshank. 191pp. 8¼ x 11. 23497-5 Pa. $6.00

AMERICAN WILDLIFE AND PLANTS, A. C. Martin, et al. Describes food habits of more than 1000 species of mammals, birds, fish. Special treatment of important food plants. Over 300 illustrations. 500pp. 5⅜ x 8½. 20793-5 Pa. $4.95

THE PEOPLE CALLED SHAKERS, Edward D. Andrews. Lifetime of research, definitive study of Shakers: origins, beliefs, practices, dances, social organization, furniture and crafts, impact on 19th-century USA, present heritage. Indispensable to student of American history, collector. 33 illustrations. 351pp. 5⅜ x 8½. 21081-2 Pa. $4.50

OLD NEW YORK IN EARLY PHOTOGRAPHS, Mary Black. New York City as it was in 1853-1901, through 196 wonderful photographs from N.-Y. Historical Society. Great Blizzard, Lincoln's funeral procession, great buildings. 228pp. 9 x 12. 22907-6 Pa. $8.95

MR. LINCOLN'S CAMERA MAN: MATHEW BRADY, Roy Meredith. Over 300 Brady photos reproduced directly from original negatives, photos. Jackson, Webster, Grant, Lee, Carnegie, Barnum; Lincoln; Battle Smoke, Death of Rebel Sniper, Atlanta Just After Capture. Lively commentary. 368pp. 8⅜ x 11¼. 23021-X Pa. $8.95

TRAVELS OF WILLIAM BARTRAM, William Bartram. From 1773-8, Bartram explored Northern Florida, Georgia, Carolinas, and reported on wild life, plants, Indians, early settlers. Basic account for period, entertaining reading. Edited by Mark Van Doren. 13 illustrations. 141pp. 5⅜ x 8½. 20013-2 Pa. $5.00

THE GENTLEMAN AND CABINET MAKER'S DIRECTOR, Thomas Chippendale. Full reprint, 1762 style book, most influential of all time; chairs, tables, sofas, mirrors, cabinets, etc. 200 plates, plus 24 photographs of surviving pieces. 249pp. 9⅞ x 12¾. 21601-2 Pa. $7.95

AMERICAN CARRIAGES, SLEIGHS, SULKIES AND CARTS, edited by Don H. Berkebile. 168 Victorian illustrations from catalogues, trade journals, fully captioned. Useful for artists. Author is Assoc. Curator, Div. of Transportation of Smithsonian Institution. 168pp. 8½ x 9½. 23328-6 Pa. $5.00

SECOND PIATIGORSKY CUP, edited by Isaac Kashdan. One of the greatest tournament books ever produced in the English language. All 90 games of the 1966 tournament, annotated by players, most annotated by both players. Features Petrosian, Spassky, Fischer, Larsen, six others. 228pp. 5⅜ x 8½. 23572-6 Pa. $3.50

ENCYCLOPEDIA OF CARD TRICKS, revised and edited by Jean Hugard. How to perform over 600 card tricks, devised by the world's greatest magicians: impromptus, spelling tricks, key cards, using special packs, much, much more. Additional chapter on card technique. 66 illustrations. 402pp. 5⅜ x 8½. (Available in U.S. only) 21252-1 Pa. $4.95

MAGIC: STAGE ILLUSIONS, SPECIAL EFFECTS AND TRICK PHOTOGRAPHY, Albert A. Hopkins, Henry R. Evans. One of the great classics; fullest, most authorative explanation of vanishing lady, levitations, scores of other great stage effects. Also small magic, automata, stunts. 446 illustrations. 556pp. 5⅜ x 8½. 23344-8 Pa. $6.95

THE SECRETS OF HOUDINI, J. C. Cannell. Classic study of Houdini's incredible magic, exposing closely-kept professional secrets and revealing, in general terms, the whole art of stage magic. 67 illustrations. 279pp. 5⅜ x 8½. 22913-0 Pa. $4.00

HOFFMANN'S MODERN MAGIC, Professor Hoffmann. One of the best, and best-known, magicians' manuals of the past century. Hundreds of tricks from card tricks and simple sleight of hand to elaborate illusions involving construction of complicated machinery. 332 illustrations. 563pp. 5⅜ x 8½. 23623-4 Pa. $6.00

MADAME PRUNIER'S FISH COOKERY BOOK, Mme. S. B. Prunier. More than 1000 recipes from world famous Prunier's of Paris and London, specially adapted here for American kitchen. Grilled tournedos with anchovy butter, Lobster a la Bordelaise, Prunier's prized desserts, more. Glossary. 340pp. 5⅜ x 8½. (Available in U.S. only) 22679-4 Pa. $3.00

FRENCH COUNTRY COOKING FOR AMERICANS, Louis Diat. 500 easy-to-make, authentic provincial recipes compiled by former head chef at New York's Fitz-Carlton Hotel: onion soup, lamb stew, potato pie, more. 309pp. 5⅜ x 8½. 23665-X Pa. $3.95

SAUCES, FRENCH AND FAMOUS, Louis Diat. Complete book gives over 200 specific recipes: bechamel, Bordelaise, hollandaise, Cumberland, apricot, etc. Author was one of this century's finest chefs, originator of vichyssoise and many other dishes. Index. 156pp. 5⅜ x 8. 23663-3 Pa. $2.75

TOLL HOUSE TRIED AND TRUE RECIPES, Ruth Graves Wakefield. Authentic recipes from the famous Mass. restaurant: popovers, veal and ham loaf, Toll House baked beans, chocolate cake crumb pudding, much more. Many helpful hints. Nearly 700 recipes. Index. 376pp. 5⅜ x 8½. 23560-2 Pa. $4.50

TONE POEMS, SERIES II: TILL EULENSPIEGELS LUSTIGE STREICHE, ALSO SPRACH ZARATHUSTRA, AND EIN HELDENLEBEN, Richard Strauss. Three important orchestral works, including very popular *Till Eulenspiegel's Marry Pranks,* reproduced in full score from original editions. Study score. 315pp. 9⅜ x 12¼. (Available in U.S. only)
23755-9 Pa. $8.95

TONE POEMS, SERIES I: DON JUAN, TOD UND VERKLARUNG AND DON QUIXOTE, Richard Strauss. Three of the most often performed and recorded works in entire orchestral repertoire, reproduced in full score from original editions. Study score. 286pp. 9⅜ x 12¼. (Available in U.S. only)
23754-0 Pa. $7.50

11 LATE STRING QUARTETS, Franz Joseph Haydn. The form which Haydn defined and "brought to perfection." *(Grove's).* 11 string quartets in complete score, his last and his best. The first in a projected series of the complete Haydn string quartets. Reliable modern Eulenberg edition, otherwise difficult to obtain. 320pp. 8⅜ x 11¼. (Available in U.S. only)
23753-2 Pa. $7.50

FOURTH, FIFTH AND SIXTH SYMPHONIES IN FULL SCORE, Peter Ilyitch Tchaikovsky. Complete orchestral scores of Symphony No. 4 in F Minor, Op. 36; Symphony No. 5 in E Minor, Op. 64; Symphony No. 6 in B Minor, "Pathetique," Op. 74. Bretikopf & Hartel eds. Study score. 480pp. 9⅜ x 12¼.
23861-X Pa. $10.95

THE MARRIAGE OF FIGARO: COMPLETE SCORE, Wolfgang A. Mozart. Finest comic opera ever written. Full score, not to be confused with piano renderings. Peters edition. Study score. 448pp. 9⅜ x 12¼. (Available in U.S. only)
23751-6 Pa. $11.95

"IMAGE" ON THE ART AND EVOLUTION OF THE FILM, edited by Marshall Deutelbaum. Pioneering book brings together for first time 38 groundbreaking articles on early silent films from *Image* and 263 illustrations newly shot from rare prints in the collection of the International Museum of Photography. A landmark work. Index. 256pp. 8¼ x 11.
23777-X Pa. $8.95

AROUND-THE-WORLD COOKY BOOK, Lois Lintner Sumption and Marguerite Lintner Ashbrook. 373 cooky and frosting recipes from 28 countries (America, Austria, China, Russia, Italy, etc.) include Viennese kisses, rice wafers, London strips, lady fingers, hony, sugar spice, maple cookies, etc. Clear instructions. All tested. 38 drawings. 182pp. 5⅜ x 8.
23802-4 Pa. $2.50

THE ART NOUVEAU STYLE, edited by Roberta Waddell. 579 rare photographs, not available elsewhere, of works in jewelry, metalwork, glass, ceramics, textiles, architecture and furniture by 175 artists—Mucha, Seguy, Lalique, Tiffany, Gaudin, Hohlwein, Saarinen, and many others. 288pp. 8⅜ x 11¼.
23515-7 Pa. $6.95

THE AMERICAN SENATOR, Anthony Trollope. Little known, long unavailable Trollope novel on a grand scale. Here are humorous comment on American vs. English culture, and stunning portrayal of a heroine/villainess. Superb evocation of Victorian village life. 561pp. 5⅜ x 8½.
23801-6 Pa. $6.00

WAS IT MURDER? James Hilton. The author of *Lost Horizon* and *Goodbye, Mr. Chips* wrote one detective novel (under a pen-name) which was quickly forgotten and virtually lost, even at the height of Hilton's fame. This edition brings it back—a finely crafted public school puzzle resplendent with Hilton's stylish atmosphere. A thoroughly English thriller by the creator of Shangri-la. 252pp. 5⅜ x 8. (Available in U.S. only)
23774-5 Pa. $3.00

CENTRAL PARK: A PHOTOGRAPHIC GUIDE, Victor Laredo and Henry Hope Reed. 121 superb photographs show dramatic views of Central Park: Bethesda Fountain, Cleopatra's Needle, Sheep Meadow, the Blockhouse, plus people engaged in many park activities: ice skating, bike riding, etc. Captions by former Curator of Central Park, Henry Hope Reed, provide historical view, changes, etc. Also photos of N.Y. landmarks on park's periphery. 96pp. 8½ x 11. 23750-8 Pa. $4.50

NANTUCKET IN THE NINETEENTH CENTURY, Clay Lancaster. 180 rare photographs, stereographs, maps, drawings and floor plans recreate unique American island society. Authentic scenes of shipwreck, lighthouses, streets, homes are arranged in geographic sequence to provide walking-tour guide to old Nantucket existing today. Introduction, captions. 160pp. 8⅞ x 11¾. 23747-8 Pa. $6.95

STONE AND MAN: A PHOTOGRAPHIC EXPLORATION, Andreas Feininger. 106 photographs by *Life* photographer Feininger portray man's deep passion for stone through the ages. Stonehenge-like megaliths, fortified towns, sculpted marble and crumbling tenements show textures, beauties, fascination. 128pp. 9¼ x 10¾. 23756-7 Pa. $5.95

CIRCLES, A MATHEMATICAL VIEW, D. Pedoe. Fundamental aspects of college geometry, non-Euclidean geometry, and other branches of mathematics: representing circle by point. Poincare model, isoperimetric property, etc. Stimulating recreational reading. 66 figures. 96pp. 5⅝ x 8¼.
63698-4 Pa. $2.75

THE DISCOVERY OF NEPTUNE, Morton Grosser. Dramatic scientific history of the investigations leading up to the actual discovery of the eighth planet of our solar system. Lucid, well-researched book by well-known historian of science. 172pp. 5⅜ x 8½. 23726-5 Pa. $3.50

THE DEVIL'S DICTIONARY. Ambrose Bierce. Barbed, bitter, brilliant witticisms in the form of a dictionary. Best, most ferocious satire America has produced. 145pp. 5⅜ x 8½. 20487-1 Pa. $2.25

HISTORY OF BACTERIOLOGY, William Bulloch. The only comprehensive history of bacteriology from the beginnings through the 19th century. Special emphasis is given to biography-Leeuwenhoek, etc. Brief accounts of 350 bacteriologists form a separate section. No clearer, fuller study, suitable to scientists and general readers, has yet been written. 52 illustrations. 448pp. 5⅝ x 8¼. 23761-3 Pa. $6.50

THE COMPLETE NONSENSE OF EDWARD LEAR, Edward Lear. All nonsense limericks, zany alphabets, Owl and Pussycat, songs, nonsense botany, etc., illustrated by Lear. Total of 321pp. 5⅜ x 8½. (Available in U.S. only) 20167-8 Pa. $3.95

INGENIOUS MATHEMATICAL PROBLEMS AND METHODS, Louis A. Graham. Sophisticated material from Graham *Dial*, applied and pure; stresses solution methods. Logic, number theory, networks, inversions, etc. 237pp. 5⅜ x 8½. 20545-2 Pa. $4.50

BEST MATHEMATICAL PUZZLES OF SAM LOYD, edited by Martin Gardner. Bizarre, original, whimsical puzzles by America's greatest puzzler. From fabulously rare *Cyclopedia*, including famous 14-15 puzzles, the Horse of a Different Color, 115 more. Elementary math. 150 illustrations. 167pp. 5⅜ x 8½. 20498-7 Pa. $2.75

THE BASIS OF COMBINATION IN CHESS, J. du Mont. Easy-to-follow, instructive book on elements of combination play, with chapters on each piece and every powerful combination team—two knights, bishop and knight, rook and bishop, etc. 250 diagrams. 218pp. 5⅜ x 8½. (Available in U.S. only) 23644-7 Pa. $3.50

MODERN CHESS STRATEGY, Ludek Pachman. The use of the queen, the active king, exchanges, pawn play, the center, weak squares, etc. Section on rook alone worth price of the book. Stress on the moderns. Often considered the most important book on strategy. 314pp. 5⅜ x 8½. 20290-9 Pa. $4.50

LASKER'S MANUAL OF CHESS, Dr. Emanuel Lasker. Great world champion offers very thorough coverage of all aspects of chess. Combinations, position play, openings, end game, aesthetics of chess, philosophy of struggle, much more. Filled with analyzed games. 390pp. 5⅜ x 8½. 20640-8 Pa. $5.00

500 MASTER GAMES OF CHESS, S. Tartakower, J. du Mont. Vast collection of great chess games from 1798-1938, with much material nowhere else readily available. Fully annoted, arranged by opening for easier study. 664pp. 5⅜ x 8½. 23208-5 Pa. $7.50

A GUIDE TO CHESS ENDINGS, Dr. Max Euwe, David Hooper. One of the finest modern works on chess endings. Thorough analysis of the most frequently encountered endings by former world champion. 331 examples, each with diagram. 248pp. 5⅜ x 8½. 23332-4 Pa. $3.75

PRINCIPLES OF ORCHESTRATION, Nikolay Rimsky-Korsakov. Great classical orchestrator provides fundamentals of tonal resonance, progression of parts, voice and orchestra, tutti effects, much else in major document. 330pp. of musical excerpts. 489pp. 6½ x 9¼. 21266-1 Pa. **$7.50**

TRISTAN UND ISOLDE, Richard Wagner. Full orchestral score with complete instrumentation. Do not confuse with piano reduction. Commentary by Felix Mottl, great Wagnerian conductor and scholar. Study score. 655pp. 8⅛ x 11. 22915-7 Pa. $13.95

REQUIEM IN FULL SCORE, Giuseppe Verdi. Immensely popular with choral groups and music lovers. Republication of edition published by C. F. Peters, Leipzig, n. d. German frontmaker in English translation. Glossary. Text in Latin. Study score. 204pp. 9⅜ x 12¼. 23682-X Pa. $6.00

COMPLETE CHAMBER MUSIC FOR STRINGS, Felix Mendelssohn. All of Mendelssohn's chamber music: Octet, 2 Quintets, 6 Quartets, and Four Pieces for String Quartet. (Nothing with piano is included). Complete works edition (1874-7). Study score. 283 pp. 9⅜ x 12¼. 23679-X Pa. **$7.50**

POPULAR SONGS OF NINETEENTH-CENTURY AMERICA, edited by Richard Jackson. 64 most important songs: "Old Oaken Bucket," "Arkansas Traveler," "Yellow Rose of Texas," etc. Authentic original sheet music, full introduction and commentaries. 290pp. 9 x 12. 23270-0 Pa. **$7.95**

COLLECTED PIANO WORKS, Scott Joplin. Edited by Vera Brodsky Lawrence. Practically all of Joplin's piano works—rags, two-steps, marches, waltzes, etc., 51 works in all. Extensive introduction by Rudi Blesh. Total of 345pp. 9 x 12. 23106-2 Pa. $14.95

BASIC PRINCIPLES OF CLASSICAL BALLET, Agrippina Vaganova. Great Russian theoretician, teacher explains methods for teaching classical ballet; incorporates best from French, Italian, Russian schools. 118 illustrations. 175pp. 5⅜ x 8½. 22036-2 Pa. $2.50

CHINESE CHARACTERS, L. Wieger. Rich analysis of 2300 characters according to traditional systems into primitives. Historical-semantic analysis to phonetics (Classical Mandarin) and radicals. 820pp. 6⅛ x 9¼. 21321-8 Pa. $10.00

EGYPTIAN LANGUAGE: EASY LESSONS IN EGYPTIAN HIEROGLYPHICS, E. A. Wallis Budge. Foremost Egyptologist offers Egyptian grammar, explanation of hieroglyphics, many reading texts, dictionary of symbols. 246pp. 5 x 7½. (Available in U.S. only) 21394-3 Clothbd. $7.50

AN ETYMOLOGICAL DICTIONARY OF MODERN ENGLISH, Ernest Weekley. Richest, fullest work, by foremost British lexicographer. Detailed word histories. Inexhaustible. Do not confuse this with *Concise Etymological Dictionary*, which is abridged. Total of 856pp. 6½ x 9¼. 21873-2, 21874-0 Pa., Two-vol. set $12.00

A MAYA GRAMMAR, Alfred M. Tozzer. Practical, useful English-language grammar by the Harvard anthropologist who was one of the three greatest American scholars in the area of Maya culture. Phonetics, grammatical processes, syntax, more. 301pp. 5⅜ x 8½. **23465-7 Pa. $4.00**

THE JOURNAL OF HENRY D. THOREAU, edited by Bradford Torrey, F. H. Allen. Complete reprinting of 14 volumes, 1837-61, over two million words; the sourcebooks for *Walden,* etc. Definitive. All original sketches, plus 75 photographs. Introduction by Walter Harding. Total of 1804pp. 8½ x 12¼. **20312-3, 20313-1 Clothbd., Two-vol. set $70.00**

CLASSIC GHOST STORIES, Charles Dickens and others. 18 wonderful stories you've wanted to reread: "The Monkey's Paw," "The House and the Brain," "The Upper Berth," "The Signalman," "Dracula's Guest," "The Tapestried Chamber," etc. Dickens, Scott, Mary Shelley, Stoker, etc. 330pp. 5⅜ x 8½. **20735-8 Pa. $4.50**

SEVEN SCIENCE FICTION NOVELS, H. G. Wells. Full novels. *First Men in the Moon, Island of Dr. Moreau, War of the Worlds, Food of the Gods, Invisible Man, Time Machine, In the Days of the Comet.* A basic science-fiction library. 1015pp. 5⅜ x 8½. (Available in U.S. only) **20264-X Clothbd. $8.95**

ARMADALE, Wilkie Collins. Third great mystery novel by the author of *The Woman in White* and *The Moonstone.* Ingeniously plotted narrative shows an exceptional command of character, incident and mood. Original magazine version with 40 illustrations. 597pp. 5⅜ x 8½. **23429-0 Pa. $6.00**

MASTERS OF MYSTERY, H. Douglas Thomson. The first book in English (1931) devoted to history and aesthetics of detective story. Poe, Doyle, LeFanu, Dickens, many others, up to 1930. New introduction and notes by E. F. Bleiler. 288pp. 5⅜ x 8½. (Available in U.S. only) **23606-4 Pa. $4.00**

FLATLAND, E. A. Abbott. Science-fiction classic explores life of 2-D being in 3-D world. Read also as introduction to thought about hyperspace. Introduction by Banesh Hoffmann. 16 illustrations. 103pp. 5⅜ x 8½. **20001-9 Pa. $2.00**

THREE SUPERNATURAL NOVELS OF THE VICTORIAN PERIOD, edited, with an introduction, by E. F. Bleiler. Reprinted complete and unabridged, three great classics of the supernatural: *The Haunted Hotel* by Wilkie Collins, *The Haunted House at Latchford* by Mrs. J. H. Riddell, and *The Lost Stradivarious* by J. Meade Falkner. 325pp. 5⅜ x 8½. **22571-2 Pa. $4.00**

AYESHA: THE RETURN OF "SHE," H. Rider Haggard. Virtuoso sequel featuring the great mythic creation, Ayesha, in an adventure that is fully as good as the first book, *She.* Original magazine version, with 47 original illustrations by Maurice Greiffenhagen. 189pp. 6½ x 9¼. **23649-8 Pa. $3.50**

UNCLE SILAS, J. Sheridan LeFanu. Victorian Gothic mystery novel, considered by many best of period, even better than Collins or Dickens. Wonderful psychological terror. Introduction by Frederick Shroyer. 436pp. 5⅜ x 8½. 21715-9 Pa. $6.00

JURGEN, James Branch Cabell. The great erotic fantasy of the 1920's that delighted thousands, shocked thousands more. Full final text, Lane edition with 13 plates by Frank Pape. 346pp. 5⅜ x 8½. 23507-6 Pa. $4.50

THE CLAVERINGS, Anthony Trollope. Major novel, chronicling aspects of British Victorian society, personalities. Reprint of Cornhill serialization, 16 plates by M. Edwards; first reprint of full text. Introduction by Norman Donaldson. 412pp. 5⅜ x 8½. 23464-9 Pa. $5.00

KEPT IN THE DARK, Anthony Trollope. Unusual short novel about Victorian morality and abnormal psychology by the great English author. Probably the first American publication. Frontispiece by Sir John Millais. 92pp. 6½ x 9¼. 23609-9 Pa. $2.50

RALPH THE HEIR, Anthony Trollope. Forgotten tale of illegitimacy, inheritance. Master novel of Trollope's later years. Victorian country estates, clubs, Parliament, fox hunting, world of fully realized characters. Reprint of 1871 edition. 12 illustrations by F. A. Faser. 434pp. of text. 5⅜ x 8½. 23642-0 Pa. $5.00

YEKL and THE IMPORTED BRIDEGROOM AND OTHER STORIES OF THE NEW YORK GHETTO, Abraham Cahan. Film *Hester Street* based on *Yekl* (1896). Novel, other stories among first about Jewish immigrants of N.Y.'s East Side. Highly praised by W. D. Howells—Cahan "a new star of realism." New introduction by Bernard G. Richards. 240pp. 5⅜ x 8½. 22427-9 Pa. $3.50

THE HIGH PLACE, James Branch Cabell. Great fantasy writer's enchanting comedy of disenchantment set in 18th-century France. Considered by some critics to be even better than his famous *Jurgen*. 10 illustrations and numerous vignettes by noted fantasy artist Frank C. Pape. 320pp. 5⅜ x 8½. 23670-6 Pa. $4.00

ALICE'S ADVENTURES UNDER GROUND, Lewis Carroll. Facsimile of ms. Carroll gave Alice Liddell in 1864. Different in many ways from final Alice. Handlettered, illustrated by Carroll. Introduction by Martin Gardner. 128pp. 5⅜ x 8½. 21482-6 Pa. $2.50

FAVORITE ANDREW LANG FAIRY TALE BOOKS IN MANY COLORS, Andrew Lang. The four Lang favorites in a boxed set—the complete *Red, Green, Yellow* and *Blue* Fairy Books. 164 stories; 439 illustrations by Lancelot Speed, Henry Ford and G. P. Jacomb Hood. Total of about 1500pp. 5⅜ x 8½. 23407-X Boxed set, Pa. $15.95

AN AUTOBIOGRAPHY, Margaret Sanger. Exciting personal account of hard-fought battle for woman's right to birth control, against prejudice, church, law. Foremost feminist document. 504pp. 5⅜ x 8½.
20470-7 Pa. $5.50

MY BONDAGE AND MY FREEDOM, Frederick Douglass. Born as a slave, Douglass became outspoken force in antislavery movement. The best of Douglass's autobiographies. Graphic description of slave life. Introduction by P. Foner. 464pp. 5⅜ x 8½. 22457-0 Pa. $5.50

LIVING MY LIFE, Emma Goldman. Candid, no holds barred account by foremost American anarchist: her own life, anarchist movement, famous contemporaries, ideas and their impact. Struggles and confrontations in America, plus deportation to U.S.S.R. Shocking inside account of persecution of anarchists under Lenin. 13 plates. Total of 944pp. 5⅜ x 8½.
22543-7, 22544-5 Pa., Two-vol. set $12.00

LETTERS AND NOTES ON THE MANNERS, CUSTOMS AND CONDITIONS OF THE NORTH AMERICAN INDIANS, George Catlin. Classic account of life among Plains Indians: ceremonies, hunt, warfare, etc. Dover edition reproduces for first time all original paintings. 312 plates. 572pp. of text. 6⅛ x 9¼. 22118-0, 22119-9 Pa.. Two-vol. set $12.00

THE MAYA AND THEIR NEIGHBORS, edited by Clarence L. Hay, others. Synoptic view of Maya civilization in broadest sense, together with Northern, Southern neighbors. Integrates much background, valuable detail not elsewhere. Prepared by greatest scholars: Kroeber, Morley, Thompson, Spinden, Vaillant, many others. Sometimes called Tozzer Memorial Volume. 60 illustrations, linguistic map. 634pp. 5⅜ x 8½.
23510-6 Pa. $10.00

HANDBOOK OF THE INDIANS OF CALIFORNIA, A. L. Kroeber. Foremost American anthropologist offers complete ethnographic study of each group. Monumental classic. 459 illustrations, maps. 995pp. 5⅜ x 8½.
23368-5 Pa. $13.00

SHAKTI AND SHAKTA, Arthur Avalon. First book to give clear, cohesive analysis of Shakta doctrine, Shakta ritual and Kundalini Shakti (yoga). Important work by one of world's foremost students of Shaktic and Tantric thought. 732pp. 5⅜ x 8½. (Available in U.S. only)
23645-5 Pa. $7.95

AN INTRODUCTION TO THE STUDY OF THE MAYA HIEROGLYPHS, Syvanus Griswold Morley. Classic study by one of the truly great figures in hieroglyph research. Still the best introduction for the student for reading Maya hieroglyphs. New introduction by J. Eric S. Thompson. 117 illustrations. 284pp. 5⅜ x 8½. 23108-9 Pa. $4.00

A STUDY OF MAYA ART, Herbert J. Spinden. Landmark classic interprets Maya symbolism, estimates styles, covers ceramics, architecture, murals, stone carvings as artforms. Still a basic book in area. New introduction by J. Eric Thompson. Over 750 illustrations. 341pp. 8⅜ x 11¼.
21235-1 Pa. $6.95